PRENTICE-HALL INTERNATIONAL SERIES IN CHEMISTRY

STEREOCHEMISTRY AND BONDING IN INORGANIC CHEMISTRY

J. E. Fergusson

University of Canterbury
Christchurch, New Zealand

Prentice-Hall, Inc., Englewood Cliffs, New Jersey

Library of Congress Cataloging in Publication Data

FERGUSSON, J. E.
   Stereochemistry and bonding in inorganic chemistry.

   (Prentice-Hall international series in chemistry)
   Includes bibliographical references.
   1. Stereochemistry. 2. Chemical bonds.
3. Chemistry, Inorganic. I. Title.
QD481.F47   546   73-12221
ISBN 0-13-846667-X

To Beverley
Linda, Wendy, and Robyn

10 9 8 7 6 5 4 3 2 1

Printed in the United States of America

PRENTICE-HALL INTERNATIONAL, INC., *London*
PRENTICE-HALL OF AUSTRALIA, PTY. LTD., *Sydney*
PRENTICE-HALL OF CANADA, LTD., *Toronto*
PRENTICE-HALL OF INDIA PRIVATE LIMITED, *New Delhi*
PRENTICE-HALL OF JAPAN, INC., *Tokyo*

Contents

v

3   Valence Bond Theory   40

4   Pauli Exclusion Principle   58

5   Molecular Orbital Theory   79

6   Equivalent Orbitals   97

Preface

A major activity in chemistry is the study of the shape of molecular and ionic species. Chemists dealing with organic compounds have been aware for many years of the dependence of organic chemistry on the stereochemistry of carbon. A similar awareness exists in inorganic chemistry, but is less well defined, due mainly to the diverse stereochemistries that exist for the elements.

The purpose of this book is to indicate some of the ways that chemists have discussed inorganic stereochemistry. In order to achieve this aim, the book is divided into two sections. Part I is a semiquantitative discussion of theoretical stereochemistry and Part II is a systematic study of the stereochemistry of the elements. Both sections, in particular Part II, are not meant to be exhaustive in their coverage of stereochemical data. Emphasis is given to systems that involve the covalent bond and to experimental data obtained from structural methods from which bond lengths and bond angles can be derived.

If the author manages to arouse an interest on the part of the reader in the problems of stereochemistry and in investigating some of the gaps that exist in our knowledge, he will be more than satisfied.

I wish to thank my colleagues Dr. H. K. J. Powell, Dr. W. T. Robinson, Dr. G. A. Rodley, Professor J. Vaughan, and Dr. G. J. Wright for reading parts of the manuscript and offering a number of helpful criticisms. My thanks also go to Mrs. P. Christensen, Mrs. J. H. Dolby, and Mrs. J. A. Humphries for typing the manuscript. Finally, I wish to acknowledge my indebtedness to the late Professor Sir R. S. Nyholm, F.R.S. who introduced me to stereochemistry.

J. E. FERGUSSON

Christchurch, New Zealand

1

Inorganic Stereochemistry

Interest in stereochemistry now covers the whole range of elements and is resulting in a vast and ever increasing accumulation of structural data.[1]* Current theory, although it allows us to recognize broad stereochemical patterns, cannot cope with the finer details. While this situation continues, the International Union of Crystallography attempt to keep up-to-date records of the known structures[2] will remain of the utmost importance, offering, as it does, our best chance of recognizing general problems and detecting gaps in the factual stereochemistry in our files.

The present book is intended merely as a guide to the current fact and theory; it does not attempt to present a detailed survey. Part I contains a simple outline of the theoretical principles that allow us, in Part II, to describe some of the general stereochemical features now recognized. This first chapter is intended to put the whole subject in historical perspective; it summarizes the history of inorganic stereochemistry up to the mid-1930's.

* Please note that all numerical references in the text and tables are to the bibliographies at the end of each chapter.

1.1   THE REGULAR ELEMENTS

The Concept of Space in Chemistry

The realization that chemical compounds could have distinctive shapes is very old, and the need for spatial considerations in chemistry was put forward as early as 1808 by Wollaston,[3] who wrote:

> I am inclined to think, that when our views are sufficiently extended, to enable us to reason with precision concerning the proportions of elementary atoms, we shall find the arithmetical relation alone will not be sufficient to explain their mutual action, and that we shall be obliged to acquire a geometrical conception of their relative arrangement in all three dimensions of solid extension.

> If there be three particles (combined with one) they might be arranged with regularity, at the angles of an equilateral triangle in a great circle surrounding the single spherule. . . . When the number of one set of particles exceeds in the proportion of four to one, then a stable equilibrium may take place if the four particles are situated at the angles of the equilateral triangles composing a regular tetrahedron.

Stereochemical implications were present in Pasteur's work on tartaric acid[4] in 1844. They eluded Pasteur at the time, although much later, in 1860, he suggested that carbon would have a tetrahedral grouping of radicals around it.[5] Three-dimensional models of tetrahedral carbon were used by Kekulé in 1867 and by Paterno in 1869,[5] and it is likely that the graphic

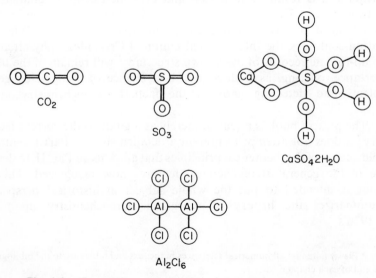

Fig. 1.1   Formulation of inorganic compounds by Crum Brown.

formulas adopted by Crum Brown in 1864 through 1866[4] (Fig. 1.1) were used with an awareness of their three-dimensional implications.

The tetrahedral stereochemistry of carbon was placed on a firm experimental foundation in two papers, published independently in 1874.[4] One, by van't Hoff, proposed that "if the four affinities of carbon are directed toward the corners of a tetrahedron of which the carbon atom itself occupies the center" then the known isomeric modifications in certain organic compounds could be accounted for. In the other paper, LeBel approached the problem by considering the consequences of an asymmetric carbon atom.

Stereochemical Studies of Other Elements

Soon after the stereochemistry of carbon had been established, a search for molecular asymmetry in compounds of other elements was undertaken. Some early reports[6] covering the period 1874–1927 are listed in Table 1.1. Since the method of investigation in these studies was almost entirely resolution of optical isomers, the findings were limited to tetrahedral, and a few octahedral, systems. It was not until techniques such as X-ray and electron diffraction, and vibrational spectroscopy became generally available that other molecular shapes were found[7] (Table 1.2).

The elucidation of the stereochemistry of nitrogen proved particularly difficult because of the problem of formulating ammonium salts. Such compounds did not appear to fit into the "fixed valency" proposal made by Kekulé, and, in terms of "variable valency" proposed by Frankland, they appeared to contain pentavalent nitrogen. In the latter case, the shape of NH_4Cl, for example, was represented in various ways (Fig. 1.2). It was Werner in 1893 who suggested that NH_4Cl could best be described as a tetrahedral arrangement of four hydrogens around the nitrogen, with the chlorine bonded by an "auxiliary valency." Werner's insight is clearly seen in the following quotation.[8]

> In the first series of the Periodic System boron, carbon and nitrogen all have coordination numbers 4, but only for carbon is this equal to the valency number. Boron, on the positive side of carbon, adds to its compounds with negative univalent elements one univalent element, and the resulting complex behaves as a univalent radical, for example $(BF_4).M$. Nitrogen, negative in regard to carbon, adds to its compounds with positive radicals one more univalent group, resulting in the formation of a positive univalent radical, as in the ammonium compounds $NR_4.X$. Thus the valency numbers of boron and nitrogen are 3 but their coordination numbers are 4. . . .

> On the other hand it may well be assumed that in the boron radical BF_4 and the ammonium radical NR_4 the four groups occupy the apices of a regular tetrahedron, as they certainly do in carbon compounds.

Table 1.1 The Stereochemistry of the Regular Elements Determined by Optical Resolution

Element	Coordination number	Stereochemistry	Workers and date[6]	Compounds
C	4	Tetrahedral	LeBel 1874 / van't Hoff 1874	Tartaric acid
N(III)	4	Tetrahedral	LeBel 1891 / Pope and Peachy 1899 / Mills and Warren 1925	$[Me(Et)Pr(iso\text{-}Bu)N]^+$ / [Benzyl(Ph)Alkyl(Me)N]I / (bicyclic ammonium structure with Br⁻)
Sn(IV)	4	Tetrahedral	Pope and Peachy 1900	$[C_2H_5(CH_3)C_3H_7SnI]$ as bromocamphorsulphonate
S(IV)	4	Tetrahedral	Pope and Peachy 1900 / Smiles 1900	$[C_2H_5(CH_3)CH_2COOHSX]$ as bromocamphorsulphonate / $[CH_3(C_2H_5)CH_2COC_6H_5SX]$ in the same way
Se(IV)	4	Tetrahedral	Pope and Neville 1902	$[(CH_3)(C_6H_5)CH_2COOHSeX]$ in the same way
Si(IV)	4	Tetrahedral	Kipping 1907, 1908	$(SO_3H \cdot C_6H_4 \cdot CH_2 - Si(C_2H_5)(C_4H_9) - O)_2$
P(V)	4	Tetrahedral	Messenheimer and Lichtenstadt 1911	$CH_3(C_2H_5)C_6H_5PO$
Te(IV)	4	Square plane / Tetrahedral	Vernon 1920, 1921 / Drew 1929	$(CH_3)_2TeI_2$ / Showed Vernon's interpretation was wrong

Ion	No.	Geometry	Reference	Structure
B(III)	4	Tetrahedral	Bueseken and Meulenhoff 1924	$\left[\begin{array}{c}\text{(benzoyl)} \,O{-}O\,B\end{array}\right]_2^{-}$ with $C{=}O$
Be(II)	4	Tetrahedral	Lowry and Burgess 1924 Mills and Gotts 1926	Be(benzoylcamphor)$_2$ Be(benzoylpyruvate)$_2$
As(V)	4	Tetrahedral	Mills and Raper 1925	[CH$_3$(C$_2$H$_5$)C$_6$H$_4$COOHAsS]
As(V)	6	Octahedral	Rosenheim and Plato 1925	$\left[\text{As}\right]_3^{-}$
S(IV)	3	Pyramidal	Phillips 1925	O—Et, O—S
			Harrison, Kenyon, and Phillips 1926	C$_6$H$_4$CH$_3$, C$_6$H$_4$COOH, O—S
			Clarke, Kenyon, and Phillips 1927	C$_6$H$_4$COOH, CH$_3$, C$_6$H$_4$COOH
			(predicted by Werner 1904)	C$_7$H$_7$SO$_2\cdot$N—S, CH$_3$
Zn(II)	4	Tetrahedral	Mills and Gotts 1926	Zn(benzoylpyruvate)$_2$
Al(III)	6	Octahedral	Wahl 1927	$\left[\text{Al}\begin{array}{c}O{-}C{=}O\\O{-}C{=}O\end{array}\right]_3^{3-}$

Table 1.2 Some Stereochemical Findings for Regular Elements Prior to 1932 Using Other than Optical Resolution

Shape	Compound	Element	Method*
Octahedral	$(NH_4)_2SiF_6$	Si	X
	K_2SnCl_6	Sn(IV)	X
Tetrahedral	CH_4	C	V, D
	CCl_4	C	V, X, E, D
	Me_4NBr	N	X
	$SiCl_4$	Si	E, D
	$GeCl_4$	Ge(IV)	E
	$SnCl_4$	Sn(IV)	E, D
	SnI_4	Sn(IV)	X
	$[Zn(CN)_4]^{2-}$	Zn	X
	$[Cd(CN)_4]^{2-}$	Cd	X
	$[Hg(CN)_4]^{2-}$	Hg(II)	X
	ClO_4^-	Cl	X
	$[O(Be_4(acetate)_6]$	O	X
Trigonal plane	CO_3^{2-}	C	X
	NO_3^-	N	X
Pyramidal	NH_3	N	V, D
	PH_3, PCl_3	P(III)	D
	$AsH_3, AsCl_3$	As(III)	D
	$SbCl_3$	Sb(III)	D
	ClO_3^-	Cl	X
	BrO_3^-	Br	X
Bent	H_2O	O	V, D
	H_2S, SO_2	S	V, D
Linear	CO_2, CS_2	C	V, X, E, D
	HCN	C	V
	ICl_2^-	I	X
	C_2H_2	C	V
	N_2O	N	V, D
	N_3^-	N	X

* X = X-ray crystal structure.
 V = Vibrational spectra.
 D = Dipole moment.
 E = Electron diffraction.

The convincing evidence for tetrahedral nitrogen in such compounds comes from the X-ray structural work of Mills and Warren (1925) and Wyckoff (1928).

In 1890, Bischoff suggested that three-coordinate nitrogen, in molecules like NR_3, would be planar. This was rejected in the same year by Hantzsch and Werner, who considered that the nitrogen in such compounds should be compared with a CH group in the compound CHR_3; NR_3 molecules should

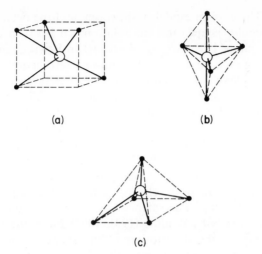

Fig. 1.2   Suggested shapes for NH_4Cl: (a) van't Hoff, 1878; (b) Willgerodt, 1888; (c) Bischoff, 1890.

therefore be pyramidal. The stereochemistry of oximes could be similarly explained and the isomers predicted[9]; for example,

$$X-C-Y \qquad\qquad X-C-Y$$
$$\parallel \qquad\text{or}\qquad \parallel$$
$$N-Z \qquad\qquad Z-N$$

were considered to be comparable to

$$X-C-Y \qquad\qquad X-C-Y$$
$$\parallel \qquad\text{or}\qquad \parallel$$
$$H-C-Z \qquad\qquad Z-C-H$$

The image of a three-coordinate nitrogen as a group with geometry similar to that of a CH group is an interesting forerunner of the picture of an NR_3 molecule in which the lone pair influences the molecular geometry, as described by Sidgwick and Powell in 1940.

By 1932, Sidgwick[7] was able to write that for a covalence (coordination number) of six the stereochemistry was invariably octahedral (so far as we know), whereas for a covalence of four the stereochemistry was tetrahedral.

Theoretical Developments

Our theoretical understanding of molecular shape began with the concept of the electron-pair bond proposed by Lewis (1916) and by Kossel (1916), and the subsequent development of this concept by Langmuir

(1919, 1920). The cubical model of the atom, and the octet rule, provided a basis for molecular shape, and the relationship between atomic structure and molecular shape became even clearer when Lewis abandoned the static cubical model of an atom for the electron-pair tetrahedral model.

The principle of isosterism proposed by Langmuir in 1919 was useful, for a time, in the assessment of molecular geometry. The principle, which is related to the law of isomorphism enunciated by Mitscherlich in 1818, states that compounds with the same number of atoms and valence electrons will tend to have the same shape. For example, carbon dioxide was known to be linear $O=C=O$, and Langmuir pointed out that the two triatomic iso-electronic species N_2O and N_3^- would probably be linear also.

The growth of quantum chemistry in the 1920's put stereochemical theory on a firm footing. The valence bond concept of directional bonding, developed particularly by Pauling (1931, 1932) and Slater (1931), proved to be a major tool in the description of the shapes of molecules. Indeed, quantum chemistry, together with the development of new structural techniques, has doubtless been the main impetus behind the growth in stereochemical studies over the last 40 years.

More Recent Times

In their Bakerian Lecture of 1940,[10] Sidgwick and Powell introduced a simple and reliable approach to stereochemistry. They showed that the broad features of inorganic stereochemistry could be rationalized by assuming that all bonding and nonbonding electron pairs around the central atom in a molecule were distributed, according to certain rules, to give minimum repulsion. This work was brought to the notice of chemists by Nyholm and Gillespie in 1957[11] and 1958,[12] and the ideas have since been developed in more detail by Gillespie. The approach will be discussed in more detail in Part I.

1.2   THE TRANSITION ELEMENTS

Early Concepts

Prior to the work of Werner, the shape of transition metal compounds had not been considered, although there had been attempts to arrive at graphic formulas. One such attempt was the chain formulation,[13,14] based on (a) the chain structures common in organic compounds, and (b) the fact that elements may exhibit variable valency. The compound

$[Co(NH_3)_6]Cl_3$, for example, was at first represented as

$$Co \begin{array}{l} \diagup NH_3 - NH_3 - Cl \\ - NH_3 - NH_3 - Cl \\ \diagdown NH_3 - NH_3 - Cl \end{array}$$

From 1875 to 1907, S. M. Jørgensen championed the chain theory, and he reformulated this hexa-ammine as

$$Co \begin{array}{l} \diagup NH_3 - Cl \\ - NH_3 - NH_3 - NH_3 - NH_3 - Cl \\ \diagdown NH_3 - Cl \end{array}$$

to explain both the existence of compounds such as $[Co(NH_3)_5Cl]Cl_2$ and $[Co(NH_3)_4Cl_2]Cl$ and their conductivities in solution.

The Co—Cl bonds, unlike the NH_3—Cl bonds, do not ionize in solution. This theory, which did not help the development of coordination chemistry, sometimes required strange formulations; for example, $[Co(NH_3)_6][Co(NO_2)_6]$[13] was represented as

$$Co \begin{array}{l} \diagup NH_3 - NO_2 - NO_2 - NO_2 - NO_2 \diagdown \\ \!\!\!\!\!\!\!\!- NH_3 \!\!-\!\!-\!\!-\!\! NO_2 \!\!-\!\!-\!\!-\!\! Co \\ \diagdown NH_3 - NH_3 - NH_3 - NH_3 - NO_2 \diagup \end{array}$$

Werner's Contribution

Werner, first in 1891 and then in more detail in 1893, rejected the chain theory and introduced both the concept of stereochemistry and "affinities" to coordination chemistry.[8] He suggested that metal ions could have two types of valencies, primary (or ionizable) and secondary (or nonionizable). The secondary valencies, usually 6 or 4 in number, could be satisfied by negative ions and/or neutral molecules. Werner also proposed that the secondary valencies were distributed in a symmetrical way around the metal ion in order to give least repulsion between the coordinated groups.

Werner, like Jørgensen, devoted much of his time to finding evidence for his theory. The evidence was principally derived from conductivity data and the resolution of optical isomers. Six-coordinate compounds, like $[Co(en)_3]Cl_3$, should exist as optical isomers, and Werner's resolution of this particular compound in 1912 and of $[Co(en)_2Cl(NH_3)]^{2+}$, $[Co(en)_2Cl_2]^+$, and $[Co(en)_2(NO_2)_2]^+$ in 1911 was convincing evidence for such stereochemical predictions. In 1914 he successfully overcame the remaining criticisms that the optical activity may be a property of carbon in the ethylene-diamine, by preparing and resolving a fully inorganic complex

$$\left[Co \left\{ \begin{array}{c} OH \\ \diagdown \\ \diagup \\ OH \end{array} Co(NH_3)_4 \right\}_3 \right] Br_6$$

The two theories predicted different conductivities for the compounds $[Co(NH_3)_3X_3]$, as follows:

$$\begin{array}{c} X \\ \diagup \\ Co-NH_3-NH_3-NH_3-X \\ \diagdown \\ X \end{array}$$

Chain theory
1-1 electrolyte

Werner's theory
nonelectrolyte

Werner used the fact that $[Co(NH_3)_3(NO_2)_3]$ had been shown to be a nonelectrolyte, by Gibbs in 1875,[14] to support his views in his paper of 1893.[8]

The chemistry of four coordinate platinum(II) was also discussed by Werner in this paper.[8] The existence of two isomers of $[PtCl_2(NH_3)_2]$ suggested to him a square planar shape for this compound, since this shape allowed cis-trans isomerism, not possible in a tetrahedral stereochemistry.

Werner's influence on coordination chemistry is unsurpassed, and his theory opened the way to new, exciting experimental work. It was once said that his contribution was so great that he left nothing for others to do. In fact, it provided a fruitful stimulus to further work, and the development in physical techniques and in quantum chemistry in the 1920's gave chemists a fresh opportunity to extend and exploit the "Werner complexes."

Later Developments in the Stereochemistry of the Transition Metals

Investigations of the stereochemistry of the transition metals up to 1930 parallels the search recorded for the regular elements. Table 1.3 gives a brief outline. Sidgwick[7] could again make the point, in 1932, that except for some square planar complexes of nickel(II), palladium(II), and platinum(II), the stereochemistry of the transition metals, where known, was

Table 1.3 The Stereochemistry of the Transition Elements

Element	Coordination number	Stereochemistry	Workers and date	Method
Cr(III)	6	Octahedron	Werner	Resolution of [Cr(en)₂Cl₂]⁺, [Cr(C₂O₄)₃]³⁻
Co(III)	6	Octahedron	Predicted, Werner 1893	Resolution of [CoCl(NH₃)(en)₂]²⁺, [CoCl(NO₂)(en)₂]⁺ and [Co(en)₃]³⁺
Fe(II)	6	Octahedron		Resolution of [Fe(bipy)₃]²⁺
Rh(III)	6	Octahedron		
Ir(III)	6	Octahedron	Delépine 1917	Resolution of [Ir(C₂O₄)₃]³⁻
Pt(IV)	6	Octahedron	Chernyaev 1928	Resolution of [Pt(NH₃)(NO₂)Cl₂(en)]⁺ and similar complex ions
			Wyckoff 1921	X-ray structure (NH₄)₂PtCl₆
Pt(II)	4	Square plane	Werner 1893	[Pt(NH₃)₂Cl₂] cis, trans isomers
			Dickinson 1922	X-ray structure K₂PtCl₄
			Cox 1932	X-ray structure [Pt(NH₃)₄]Cl₂H₂O
Pd(II)	4	Square plane	Dickinson 1922	X-ray structure K₂PdCl₄, (NH₄)₂PdCl₄
Cu(II)	4	Tetrahedral	Mills and Gotts 1926	Resolution of Cu(benzoylpyruvate)₂, may be in doubt
	4	Square plane	Hendricks and Dickinson 1927	X-ray structure of K₂CuCl₄2H₂O, very distorted octahedron
	6	Octahedron	Cox and Webster 1935	X-ray structure of Cu(II) β-diketone complexes
	4	Square plane	Wahl 1928	Resolution of [Cu(en)₂(H₂O)₂]²⁺
Ni(II)	4	Square plane	Sugden 1932	cis, trans isomers of bis(benzylmethylglyoxime)nickel(II)
	6	Octahedral	Wahl 1928	Resolution of [Ni(en)₂(H₂O)₂]²⁺
			Morgan and Burstall 1931	Resolution of [Ni(bipy)₃]²⁺
V(V)	4	Tetrahedral		X-ray structure of VO₄³⁻ salts
Mn(VII)	4	Tetrahedral		X-ray structure of KMnO₄

11

either octahedral or tetrahedral. The problem of distorted octahedral complexes of copper (Table 1.3) was not noticed until later.

The stereochemistry on which it was most difficult to obtain evidence was the square plane.[7] In certain compounds of the type Pta_2b_2 it was clear that the "isomers" were, in fact, monomeric Pta_2b_2 and dimeric $Pt_2a_4b_4$ compounds. Reihlen (1926) demonstrated that the properties of the two forms of $[Pt(NH_3)_2Cl_2]$ (the compound mentioned by Werner in 1893) were consistent with monomeric and dimeric structures. However, Hantzsch (1926) showed that the two forms of $[PtCl_2(py)_2]$ were true cis-trans isomers. Finally, X-ray structural analyses confirmed Werner's square planar structures. (See Table 1.3.)

Theory of Stereochemistry

The next major step came with the work of Pauling (1931, 1932),[15] who proposed that the shapes of coordination compounds could be understood in terms of hybridized metal orbitals. This suggestion, together with the concept of the dative covalent bond proposed by Sidgwick (1923, 1927), enabled Pauling to account for the magnetic properties of the transition metals as well as their stereochemistry. For example, if nickel forms square planar complexes then it must be diamagnetic, because only complete spin pairing can provide the d-orbital required for square planar dsp^2 hybrids.

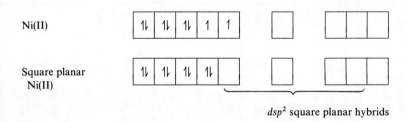

dsp^2 square planar hybrids

Octahedral spin-paired cobalt(III) compounds are also adequately explained by Pauling's theory.

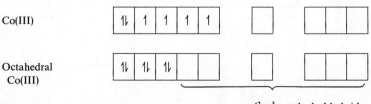

d^2sp^3 octahedral hybrids

With this theory, it is difficult to explain octahedral spin-free iron(II) compounds (unless outer d-orbitals are invoked in bonding) and the existence of square planar rather than tetrahedral copper(II) complexes. Irrespective of these problems, the influence of Pauling's work in the development of coordination chemistry was considerable.

Recent Developments

Crystal field theory was developed by Bethe (1929), Schlapp and Penny (1932), and Van Vleck (1935). Its introduction to chemists by Hartmann and Schlafer (1951) and Orgel (1952) emphasized the importance of stereochemistry in the interpretation of the physical and chemical properties of transition metal complexes. The impetus of this theory and, more recently, of the molecular orbital theory, together with the wide range of structural techniques available for the study of transition metal complexes, has resulted in a considerable increase in the available stereochemical data.

1.3   TEXTBOOKS ON STEREOCHEMISTRY

In most areas of chemistry, the appearance of key textbooks has been of as much significance as research work in the field. This is also the case in inorganic stereochemistry. The following books gave timely, authoritative, and stimulating summaries of the state of the field up to their publication date:

A. WERNER, *New Ideas on Inorganic Chemistry*, 1905 and German editions. English translation by E. P. Hedley, Longmans, Green & Company, Ltd., London, 1911.

N. V. SIDGWICK, *The Electronic Theory of Valency*, Clarendon Press, Oxford, 1927.

N. V. SIDGWICK, *The Covalent Link in Chemistry*, Cornell University Press, Ithaca, N. Y., 1933.

G. T. MORGAN and F. H. BURSTALL, *Inorganic Chemistry*, W. Heffer and Sons, Ltd., Cambridge, Engl., 1936.

W. HÜCKEL, *Structural Chemistry of Inorganic Chemistry*, Vols. 1 and 2, English translation by L. H. Long, Elsevier Publishing Company, Inc., Amsterdam, 1950.

H. J. EMELEUS and J. S. ANDERSON, *Modern Aspects of Inorganic Chemistry*, Routledge and Kegan Paul, Ltd., London, 1938. Subsequent editions, 1952 and 1960.

L. PAULING, *The Nature of the Chemical Bond*, Cornell University Press, Ithaca, N. Y., 1938. Subsequent editions, 1940, 1960.

A. F. WELLS, *Structural Inorganic Chemistry*, Clarendon Press, Oxford, 1945. Subsequent editions, 1950 and 1962.

W. KLYNE, ed., *Progress in Stereochemistry*, Vol. 1, Butterworths, London, 1954. See papers by A. D. WALSH (p. 1), R. S. NYHOLM (p. 322), A. MACCOLL (p. 361).

W. KLYNE and P. B. D. DE LA MARE, eds., *Progress in Stereochemistry*, Vol. 2, Butterworths, London, 1958. See paper by R. J. GILLESPIE and R. S. NYHOLM, p. 261.

These books are a valuable source of information on the development of inorganic stereochemistry.

1.4   PHYSICAL METHODS OF STRUCTURE DETERMINATION

Table 1.4 lists the physico-chemical methods that give information on molecular structure. The first four methods give accurate structural data and in many cases complete structural information. X-ray diffraction of crystalline compounds is among the best and most accurate of the methods available.[16] This is due in part to the availability of fast computers, the automation of data collection, from which chemists can glean improved data, and the exact mathematical relationship between diffraction pattern and structure.

The remaining methods give information on the general and broad features of stereochemistry—whether, for example, the stereochemistry of an atom in a molecule is octahedral or tetrahedral. Finer details, such as distortions from a regular shape, are only obtained from some of the methods by carrying out accurate detailed studies.

The choice of technique used in solving a stereochemical problem depends on the information sought. It is often necessary, however, to be cautious about interpretation of the available facts, particularly for compounds where two or more shapes are possible, all of which are comparable in energy.

Another factor to consider when deducing molecular structure from physical data is the time scale of the structural technique used. For example, a change in a molecule that occurs in less than 10^{-8} sec will not be observed when the nuclear magnetic resonance (nmr) experiment is carried out. In the case of nmr, useful structural information may, in fact, be obtained by altering the rate of molecular changes by variation of the temperature. A list of the time constants for a number of structural methods is given in Table 1.5.

The discussion in the following chapters is mainly based on data available from the first four techniques listed in Table 1.4. Details on all the methods, the background theory and relevant applications, are readily available in the literature, and a few standard sources are listed in the bibliography.[17]

Table 1.4 Methods of Obtaining Molecular Structure

Method	Requirements	Results obtained
X-ray diffraction	Crystalline compounds, X-ray source and means of collecting data, computer.	Complete structural information, internuclear distances and angles to a high degree of accuracy.
Electron diffraction	Compound in gaseous state, though crystalline and liquid state can be studied.	Complete structural information, cannot deal with too many structural parameters.
Neutron diffraction	Crystalline sample (large size) high neutron flux beam.	Complete structural data most useful for finding light atoms, distinguishing atoms of similar atomic number (both of which are problems using X-ray diffraction) and studying ferro- and anti-ferromagnetic materials.
Microwave spectroscopy	Gaseous material at a vapor pressure of at least 10^{-3} mm, compound must have a dipole moment.	Structural data, bond lengths for simple (molecular) compounds.
Infrared and Raman spectroscopy	Gaseous, liquid or crystalline samples.	Molecular shape based on symmetry of the central atom. Bond lengths from accurate vibration-rotation spectra.
Ultraviolet-visible spectroscopy	Especially for transition metal complexes, in all states. Single crystal study most accurate.	Symmetry of ligand environment (broad features only).

Table 1.4 (*cont.*)

Method	Requirements	Results obtained
Nuclear magnetic resonance spectroscopy	Usually diamagnetic compounds soluble in a solvent.	Gives the environment of atoms and groups in a molecule.
Optical resolution, optical rotatory dispersion (ORD), and circular dichroism (CD)	Solid compounds, soluble in suitable solvents. For ORD and CD need to investigate the ultraviolet-visible spectral region.	Optical activity, hence presence of asymmetry and probable stereochemistry.
Mass spectrometry	Gaseous or reasonably volatile compound.	Fragmentation pattern can be a guide to molecular shape.
Mössbauer spectroscopy	For particular elements, e.g., Fe, Sn, Ru, Te; γ source.	Gives the symmetry or environment of the Mössbauer nucleus.
Nuclear quadrupole resonance spectroscopy	Solid compound (large sample). Useful for halide compounds (Cl, Br, I).	Distinguishes different environments of the resonating nucleus in the crystalline state.
Electron spin resonance spectroscopy	Requires a paramagnetic species.	Gives information of the ligand symmetry around metal ions.
X-ray absorption spectroscopy and photo-electron spectroscopy	Mostly solid samples for X-ray absorption spectroscopy, solid or gas can be used for photo-electron spectroscopy.	Can give information on the molecular environment of atoms or ions.
Magnetic measurements	Sample as solid, or in solution.	Results can be related to the ligand symmetry around a transition metal ion.
Dipole moment	Gas, liquid or solid soluble in a suitable solvent.	A guide to molecular geometry.

Table 1.5 Time Constants for a Number of Spectroscopic Techniques

Spectroscopic technique	Time constant (sec)
Electronic ⎱ Vibrational ⎰	$> 10^{-12}$
Rotational	$\sim 10^{-10}$
ESR	$10^{-8}-10^{-10}$
NMR	10^{-8}
Mössbauer	10^{-7}

BIBLIOGRAPHY

1. N. V. A. OOSTHOEK, *Structure Reports*, International Union of Crystallography, Uitgevers, Mij., Utrecht, 1940–1962. R. W. G. WYCKOFF, *Crystal Structures*, Interscience Publishers, New York, 1960. L. E. SUTTON, ed., *Interatomic Distances, Chem. Soc. London Spec. Publ.* 11 (1958), 18 (1965); and A. F. WELLS, *Structural Inorganic Chemistry*, 3rd ed., Clarendon Press, Oxford, 1962.

2. O. KENNARD and D. C. WATSON, eds., *Molecular Structure and Dimensions*, Vols. 1 and 2, International Union of Crystallography, Cambridge, 1970.

3. W. H. WOLLASTON, *Phil. Trans. Roy. Soc. London, Ser. A*, **98**, 96 (1808).

4. W. C. PALMER, *A History of the Concept of Valency to 1930*, Cambridge University Press, Cambridge, 1965.

5. A. FINDLEY, *A Hundred Years of Chemistry*, 2nd ed., Gerald Duckworth & Co., Ltd., London, 1948.

6. N. V. SIDGWICK, *The Electronic Theory of Valency*, Clarendon Press, Oxford, 1927.

7. N. V. SIDGWICK, *The Covalent Link in Chemistry*, Cornell University Press, Ithaca, N. Y., 1933.

8. A. WERNER, *Z. Anorg. Chem.*, **3**, 267 (1893).

9. G. B. KAUFFMAN, ed., *Werner Centennial, Adv. Chem. Series*, No. 62, ed. by G. B. Kauffman, 1967, p. 41.

10. N. V. SIDGWICK and H. M. POWELL, *Proc. Roy. Soc. London, Ser. A*, **176**, 153 (1940).

11. R. J. GILLESPIE and R. S. NYHOLM, *Quart. Rev. London*, **11**, 339 (1957).

12. R. J. GILLESPIE and R. S. NYHOLM, *Progr. Stereochem.*, **2**, 261 (1958).

13. P. S. COHEN, *Werner Centennial, Adv. Chem. Series*, No. 62, ed. by G. B. Kauffman, 1967, p. 8; G. B. KAUFFMAN, *Chymia*, **6**, 180 (1960).

14. *Fasciculus Extraordinarius, Alfred Werner (1866–1919)*, *Helv. Chim. Acta*, Basel, 1967; see article by C. Schwarzenbach, p. 38.

15. L. PAULING, *J. Am. Chem. Soc.*, **53**, 1367, 3225 (1931).

16. P. P. EWALD, ed., *Fifty Years of X-ray Diffraction*, International Union of Crystallography, Utrecht, 1962.

17. H. A. O. HILL and P. DAY, eds., *Physical Methods in Advanced Inorganic Chemistry*, Interscience Publishers, New York, 1968; W. L. JOLLY, *The Synthesis and Characterization of Inorganic Compounds*, Prentice-Hall, Inc., Englewood Cliffs, N. J., 1970; R. S. DRAGO, *Physical Methods in Inorganic Chemistry*, Reinhold Publishing Corp., New York, 1965; and H. B. JONASSEN and A. WEISSBERGER, eds., *Technique of Inorganic Chemistry*, Vols. 1–8, Interscience Publishers, New York, 1963–1968.

PART I

2
Atomic Orbitals

2.1 WAVE EQUATION

The motion of an electron associated with a nucleus may be described in terms of a wave equation. The equation can be solved either exactly or approximately to give three important results: the most probable distance of the electron from the nucleus; its most probable position in space; and the energy of the different electronic states. The wave equation is exactly soluble for only a one-electron system (hydrogen atom), but fortunately the results obtained can be used with some modification for many-electron atoms.

For the one-electron system, the relation between the amplitude of the wave ψ_{xyz} and the total energy of the system E is given by the Schrödinger equation[1]

$$\nabla^2 \psi_{xyz} + \frac{8\pi^2 m}{h^2}(E - V)\psi_{xyz} = 0 \tag{2.1}$$

The equation can be solved for hydrogen by first converting to polar coordinates, then separating the variables r, θ, and ϕ, and finally solving the three separate equations.

The combined result of the three separate solutions is given by the equation[1]

$$\psi_{nlm} = R_{nl}(r)\Theta_{lm}(\theta)\Phi_m(\phi) \tag{2.2}$$

or

$$\psi_{nlm} = Ce^{-Zr/na_o}\left(\frac{2Zr}{na_o}\right)^l L_{n+l}^{2l+1} \frac{2Zr}{na_o} P_l^{|m|} \cos\theta\, e^{\pm im\phi} \tag{2.3}$$

where C is a constant and a_o, the Bohr radius (53 pm), corresponds to the first atomic orbit of the Bohr hydrogen atom. The terms L_{n+l}^{2l+1} and $P_l^{|m|}$ are the associated Laguerre and Legendre polynomials, respectively.

The amplitude ψ_{nlm} is called the wave function, the square of which (ψ_{nlm}^2 or $\psi_{nlm}\psi_{nlm}^*$) is a measure of the probability of finding the electron at some point in space. The wave function ψ_{nlm} is dependent on the quantum numbers n, l, and m, which arise from solutions of three separate equations in $R(r)$, $\Theta(\theta)$, and $\Phi(\phi)$. The principal quantum number n has positive integral values, $1, 2, 3, \ldots, l$, the azimuthal or angular momentum quantum number has values $0, 1, \ldots, n-1$, and m, the magnetic quantum number, has the values $0, \pm 1, \pm 2, \ldots, \pm l$. The latter quantum number determines the orientation of the wave function in space in the presence of an external magnetic field.

2.2 ORBITALS

The wave function ψ_{nlm}, in the case of an atom, is termed an atomic orbital. The one-electron orbitals are completely defined by the symbol nl_m and are labeled $s, p, d, f, \ldots$, orbitals according to the values of $l = 0, 1, 2, 3, \ldots$, respectively (Table 2.1).

Table 2.1

n	l	m	Orbital
1	0	0	$1s$ Orbital
2	0	0	$2s$ Orbital
	1	$0, \pm 1$	$2p_{0,\pm 1}$ Orbitals
3	0	0	$3s$ Orbital
	1	$0, \pm 1$	$3p_{0,\pm 1}$ Orbitals
	2	$0, \pm 1, \pm 2$	$3d_{0,\pm 1,\pm 2}$ Orbitals
4	0	0	$4s$ Orbital
	1	$0, \pm 1$	$4p_{0,\pm 1}$ Orbitals
	2	$0, \pm 1, \pm 2$	$4d_{0,\pm 1,\pm 2}$ Orbitals
	3	$0, \pm 1, \pm 2, \pm 3$	$4f_{0,\pm 1,\pm 2,\pm 3}$ Orbitals

The solutions of the Schrödinger Equation (2.3) for the hydrogen atom (in their present form) are of little practical value to the chemist interested in the stereochemical properties of atomic orbitals. The angular terms of the wave functions $Y_{lm} = \Theta_{lm}(\theta)\Phi_m(\phi)$ determine the shape of the orbitals and the direction that they point in space. Except when $m = 0$, the angular wave functions consist of real and imaginary parts, due to the form of the function $\Phi_m(\phi)$, i.e.

$$\Phi_m(\phi) = e^{\pm im\phi}$$

$$= \cos(m\phi) \pm i \sin(m\phi) \tag{2.4}$$

To obtain real functions, we take linear combinations of the wave functions so that the imaginary terms cancel. The new functions are valid solutions of the Schrödinger equation, because any linear combination of solutions is itself a solution. However, we must ensure that the new functions are normalized and orthogonal, i.e., they conform to the relationships $\int (\psi_{nlm})^2 \, dT = 1$ and $\int (\psi'_{nlm})(\psi''_{nlm}) \, dT = 0$. The most common linear combinations are

and

$$\frac{1}{\sqrt{2}}[Y_{l(+m)} + Y_{l(-m)}]$$

$$-\frac{i}{\sqrt{2}}[Y_{l(+m)} - Y_{l(-m)}] \tag{2.5}$$

for all $m \neq 0$.

For example, for p orbitals we have the linear combinations (2.6).

$$\frac{1}{\sqrt{2}}(Y_{11} + Y_{1-1}) = \left(\frac{\sqrt{3}}{2\sqrt{\pi}}\right) \sin\theta \cos\phi = \left(\frac{\sqrt{3}}{2\sqrt{\pi}}\right)\frac{x}{r}; \quad p_x$$

$$-\frac{i}{\sqrt{2}}(Y_{11} - Y_{1-1}) = \left(\frac{\sqrt{3}}{2\sqrt{\pi}}\right) \sin\theta \sin\phi = \left(\frac{\sqrt{3}}{2\sqrt{\pi}}\right)\frac{y}{r}; \quad p_y \tag{2.6}$$

$$Y_{10} = \left(\frac{\sqrt{3}}{2\sqrt{\pi}}\right) \cos\theta = \left(\frac{\sqrt{3}}{2\sqrt{\pi}}\right)\frac{z}{r}; \quad p_z$$

The new functions for the s, p, and d orbitals, up to $n = 3$, in terms of both polar and Cartesian coordinates, are listed in Table 2.2. It is clear from the expression for the orbitals in Cartesian coordinates why the various subscripts are used for labeling the atomic orbitals.

The pictorial representation of the angular functions for the 1s, 2p, and 3d orbitals are given in Fig. 2.1. It is apparent from these diagrams that the 2p and 3d orbitals have prominent directional properties, which are the basis of stereochemical theories for chemical compounds. However, remember that the particular linear combination previously described is

Table 2.2

n	l	m	$R_{nl}(r)$ (Radial term)	$Y_{lm}(\theta\phi)$ (Angular terms)	$Y_{lm}(\theta\phi)$ (in terms of Cartesian coordinates)	Orbital
1	0	0	$2\left(\dfrac{Z}{a_o}\right)^{3/2} e^{-\sigma/2}*$	$\left(\dfrac{1}{4\pi}\right)^{1/2}$	$\left(\dfrac{1}{4\pi}\right)^{1/2}$	$1s$
2	0	0	$\dfrac{1}{2\sqrt{2}}\left(\dfrac{Z}{2a_o}\right)^{3/2}(2-\sigma)e^{-\sigma/2}$	$\left(\dfrac{1}{4\pi}\right)^{1/2}$	$\left(\dfrac{1}{4\pi}\right)^{1/2}$	$2s$
2	1	0	$\dfrac{1}{2\sqrt{6}}\left(\dfrac{Z}{a_o}\right)^{3/2}\sigma e^{-\sigma/2}$	$\left(\dfrac{3}{4\pi}\right)^{1/2}\cos\theta$	$\left(\dfrac{3}{4\pi}\right)^{1/2}\dfrac{z}{r}$	$2p_z$
2	1	$\left.\begin{array}{c} \\ \\ \end{array}\right\}\pm1$		$\left(\dfrac{3}{4\pi}\right)^{1/2}\sin\theta\sin\phi$	$\left(\dfrac{3}{4\pi}\right)^{1/2}\dfrac{y}{r}$	$2p_y$
2	1			$\left(\dfrac{3}{4\pi}\right)^{1/2}\sin\theta\cos\phi$	$\left(\dfrac{3}{4\pi}\right)^{1/2}\dfrac{x}{r}$	$2p_x$
3	0	0	$\dfrac{1}{9\sqrt{3}}\left(\dfrac{Z}{a_o}\right)^{3/2}(6-6\sigma+\sigma^2)e^{-\sigma/2}$	$\left(\dfrac{1}{4\pi}\right)^{1/2}$	$\left(\dfrac{1}{4\pi}\right)^{1/2}$	$3s$

3	1	0	$\dfrac{1}{9\sqrt{6}}\left(\dfrac{Z}{a_o}\right)^{3/2}(4-\sigma)e^{-\sigma/2}$	$\left(\dfrac{3}{4\pi}\right)^{1/2}\cos\theta$	$\left(\dfrac{3}{4\pi}\right)^{1/2}\dfrac{z}{r}$	$3p_z$
3	1	$\left.\begin{array}{c}\pm 1 \\ \pm 1\end{array}\right\}$		$\left(\dfrac{3}{4\pi}\right)^{1/2}\sin\theta\sin\phi$	$\left(\dfrac{3}{4\pi}\right)^{1/2}\dfrac{y}{r}$	$3p_y$
3	1			$\left(\dfrac{3}{4\pi}\right)^{1/2}\sin\theta\cos\phi$	$\left(\dfrac{3}{4\pi}\right)^{1/2}\dfrac{x}{r}$	$3p_x$
3	2	0	$\dfrac{1}{9\sqrt{30}}\left(\dfrac{Z}{a_o}\right)^{3/2}\sigma^2 e^{-\sigma/2}$	$\left(\dfrac{5}{16\pi}\right)^{1/2}(3\cos^2\theta-1)$	$\left(\dfrac{5}{16\pi}\right)^{1/2}\dfrac{11}{2}(3z^2-r^2)$	$3d_{z^2}$
3	2	$\left.\begin{array}{c}\pm 1 \\ \pm 1\end{array}\right\}$		$\left(\dfrac{15}{4\pi}\right)^{1/2}\sin\theta\cos\theta\cos\phi$	$\left(\dfrac{15}{16\pi}\right)^{1/2}\sqrt{3}xz$	$3d_{xz}$
3	2			$\left(\dfrac{15}{4\pi}\right)^{1/2}\sin\theta\cos\theta\sin\phi$	$\left(\dfrac{15}{4\pi}\right)^{1/2}\sqrt{3}yz$	$3d_{yz}$
3	2	$\left.\begin{array}{c}\pm 2 \\ \pm 2\end{array}\right\}$		$\left(\dfrac{15}{4\pi}\right)^{1/2}\sin^2\theta\sin 2\phi$	$\left(\dfrac{15}{4\pi}\right)^{1/2}\sqrt{3}xy$	$3d_{xy}$
3	2			$\left(\dfrac{15}{4\pi}\right)^{1/2}\sin^2\theta\cos 2\phi$	$\left(\dfrac{15}{4\pi}\right)^{1/2}\dfrac{\sqrt{3}}{2}(x^2-y^2)$	$3d_{x^2-y^2}$

* $\sigma=\dfrac{2Zr}{na_o}$ and $a_o=\dfrac{h^2}{4\pi^2 me^2}$.

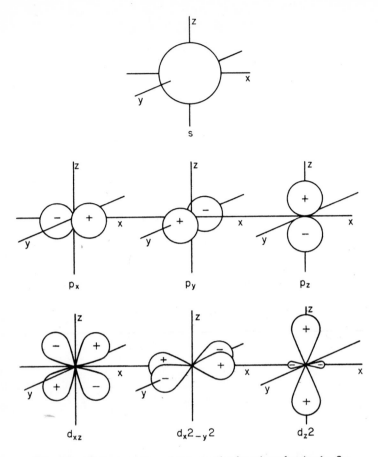

Fig. 2.1 Representation of the angular functions for the 1*s*, 2*p*, and 3*d* orbitals.

convenient for the discussion of atomic and molecular properties. Other valid linear combinations can be made that may well be more suited to different problems. The probability function $(\psi_{nlm})^2$ has identical directional properties to the wave function ψ_{nlm}. Therefore, for the purposes of discussing stereochemistry, either of these functions can be used.

Atomic orbitals of higher quantum levels also have similar directional properties. The graphical representation of the radial wave function, $R_{nl}(r)$, plotted against r (Fig. 2.2), shows that a 3*p* orbital differs from a 2*p* in having an extra node where the radial wave function crosses the axis at $R_{nl}(r) = 0$. The difference is emphasized in the diagrams of the angular wave function of a $2p_x$ and $3p_x$ orbital (Fig. 2.3). The directional properties are the same

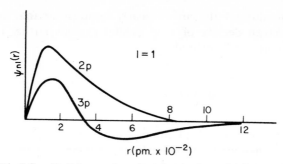

Fig. 2.2   Radial wave functions for the 2*p* and 3*p* orbitals.

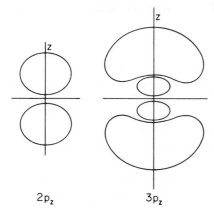

$2p_z$ $3p_z$

Fig. 2.3   Angular wave functions for the $2p_z$ and $3p_z$ orbitals.

but are more pronounced for the 3*p* orbital. This is made clear from the radial distribution curves (Fig. 2.4) $4\pi r^2 R_{nl}(r)^2$, which give the probability of finding the electron in a spherical shell of radius *r* and $r + dr$. The curve for the 3*p* orbital shows that the most probable position of the electron is in the

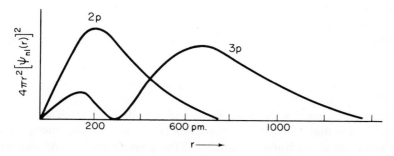

Fig. 2.4   Radial distribution curves for the 2*p* and 3*p* orbitals.

outer lobe. In addition, the curves clearly demonstrate that the position of maximum electron density of a $3p$ orbital extends farther out from the nucleus than that of a $2p$ orbital.

2.3 MANY-ELECTRON ATOMS

Atomic systems more complex than the hydrogen atom give rise to computation problems when we try to solve the Schrödinger equation. The reason for the difficulty is electron–electron repulsion, which occurs in the potential energy term. Various approximate methods of dealing with the difficulty have been used.[2,3] One of the more successful is the Hartree–Fock Self Consistent Field (SCF) method.[2,3] The type of wave function used is

$$\psi = \left(\sum_{n\xi} C_{n\xi} r^n l^{-\xi r} \right) Y_{lm}(\theta\phi) \tag{2.7}$$

Empirical rules have been developed for determining the value of ξ, which involves an estimate of the shielding constant.[2] From Equation (2.7), there is no alteration in the angular wave function; therefore, the orbitals have the same shape and directional properties as for the hydrogen atom. We are thus justified in using the same method of labeling the orbitals as described previously.

Two further conclusions are also important in the present context. Although the shape of, say, an np orbital remains the same in various atoms, its size changes with nuclear charge, because the position of maximum electron density is inversely proportional to the nuclear charge. As the number of electrons increase, the orbitals become smaller in size, since electrons become less efficient in shielding the nuclear charge as they get further from the nucleus, and there is an overall contraction in orbital size. The curves shown in Fig. 2.5 are plots of orbital size against atomic number.[4] The points on the curves are obtained from the maximum in the radial distribution curves using the SCF method.

The energy of the atomic orbitals obtained from the Schrödinger equation for a hydrogen atom is dependent on the quantum number n,

$$E = -\frac{Z^2 e^2}{2n^2 a_o} \tag{2.8}$$

For example, when $n = 3$, the $3s$, $3p$, and $3d$ orbitals all have the same energy. However, for many-electron orbitals this is not so, and the energy is also dependent on the quantum number l. This gives us our second important conclusion—that as l increases for a particular n, the energy of the various

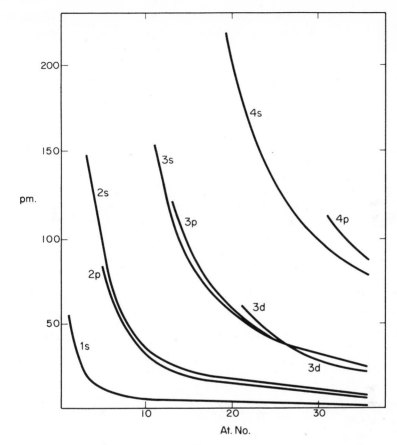

Fig. 2.5 Radius of maximum charge density plotted against atomic number for wave functions 1*s* to 4*p*.

orbitals increases. The combined effect of size decrease and energy increase has an interesting influence on the 3*d* electrons, viz., they are of comparable energy to the 4*s* electrons but the 3*d* orbitals are smaller in size.

2.4   SYMMETRY ORBITALS

It is usual to classify atomic orbitals according to the quantum numbers, *n*, *l*, and *m*. For a discussion of stereochemistry, an additional classification, in terms of the symmetry in which an atom finds itself in a molecule, is useful. We shall now describe the method of obtaining symmetry orbitals, assuming that the reader has a knowledge of group theory and use of group character tables.[5]

Molecules can be described according to a symmetry point group. The point groups are made up of different combinations of the symmetry elements listed in Table 2.3, and the point groups of greatest chemical interest are listed

Table 2.3

Symmetry elements	Comments
E	Identity.
σ	Plane of symmetry, σ_h reflection in a plane perpendicular to principal axis, σ_v reflection in a plane containing the principal axis.
C_n	Rotation axis where n signifies the order of the axis, the angle of rotation is $2\pi/n$.
S_n	Improper rotation axis, rotation by $2\pi/n$ followed by a reflection in a plane perpendicular to the rotation axis.
i	Inversion through a center of symmetry.

in Table 2.4. As the name "point group" implies, it is concerned with the symmetry around a point. The point in molecular systems generally coincides with the central atom—e.g., boron in BCl_3 and iron in $Fe(CN)_6^{3-}$.

Table 2.4

Point group	Symmetry elements
C_s	A plane of symmetry.
C_i	A center of symmetry.
C_n	n-Fold rotation axis, $n = 1 - 8$.
D_n	n-Fold axis and n horizontal twofold axes, $n = 2 - 6$.
C_{nv}	n-Fold axis, n vertical planes, $n = 2 - 6$.
C_{nh}	n-Fold axis, horizontal plane,. $n = 2 - 6$.
D_{nh}	Rotations as for D_n, one horizontal plane and n vertical planes containing the horizontal axes, $n = 2 - 6$.
D_{nd}	Rotations as for D_n, n vertical planes bisecting angles between horizontal axes, $n = 2 - 5$.
S_n	Alternating axes, $n = 4, 6, 8$.
T_d, O_h, O	Cubic groups, tetrahedron, and octahedron.
$C_{\infty v}, D_{\infty h}$	Infinite rotation axis, linear systems.
I_h, I	Icosahedron.

It is first necessary to classify a molecule in terms of its symmetry; two examples, water and phosphorus pentachloride, will be considered.

Water has a "V" shape, and we shall consider it in relation to the Cartesian axes x, y, z (Fig. 2.6). The symmetry elements of the molecule are a

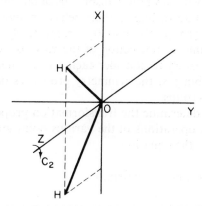

Fig. 2.6   The coordinate system for water.

C_2 axis in the plane of the atoms, bisecting the HOH angle, a reflection plane containing this axis and perpendicular to the plane of the molecule, and a reflection plane in the plane of the molecule. Hence, the point group for water is C_{2v}. Phosphorus pentachloride (gaseous) has a trigonal bipyramidal stereochemistry (Fig. 2.7). The symmetry elements are a C_3 rotation

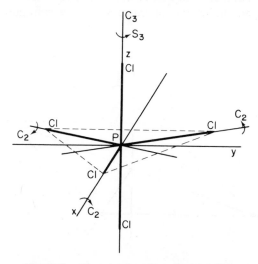

Fig. 2.7   The coordinate system for PCl_5.

axis and three C_2 axes perpendicular to the C_3 axis. These symmetry elements classify the molecule initially as belonging to a D group. In addition, there are three vertical planes of symmetry containing the C_3 axis and one horizontal plane. Therefore, according to the information in Table 2.4, the molecule has D_{3h} symmetry.

The orbitals of the oxygen of water and the phosphorus of phosphorus pentachloride can now be classified according to their behavior under the symmetry operations of the point groups C_{2v} and D_{3h}, respectively, inasmuch as these atoms lie at the point of intersection of the various axes of rotation. Fortunately, it is not necessary to consider each central chemical element in each of its compounds but just the symmetry properties of s, p, and d orbitals in the various point groups.

The basis of the method is to determine the transformation properties of the orbitals under the symmetry operations of the various point groups. This is described mathematically by the equation

$$\hat{R}\psi = (\text{transformation matrix})\psi \qquad (2.9)$$

where $\hat{R}$ is the symmetry operation and ψ the atomic wave function. The sum of the diagonal terms of the transformation matrix (the trace) gives the character χ for the operation $\hat{R}$. After consideration of all the symmetry operations of the point group, a list of characters, called a representation, allows the orbital ψ to be classified according to the representation to which it corresponds in the character table of the point group.

We shall consider in some detail three point groups: C_{2v}, D_{4h}, and D_{3h} (Figs. 2.6, 2.7, and 2.8). Because s orbitals have spherical symmetry, they remain unchanged under the operations of any symmetry point group.

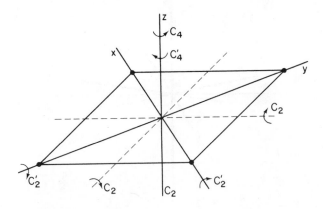

Fig. 2.8   The coordinate system for a square plane.

Therefore, we have

$$\hat{R}(\psi_s) = (1)\psi_s \tag{2.10}$$

for any $\hat{R}$ of any point group. For the preceding point groups, the s orbitals transform as the a_1, a_{1g}, and a_1' representations, respectively. It is the general practice to use lowercase letters when describing orbitals in terms of the representations of point groups. The orbital transformations for the three point groups C_{2v}, D_{4h}, and D_{3h} are tabulated in Tables 2.5, 2.6, and 2.7,

Table 2.5 Orbital Transformation (C_{2v})

Symmetry operation	p_x	p_y	p_z	d_{xy}	d_{yz}	d_{zx}	$d_{x^2-y^2}$	d_{z^2}
C_2	$-p_x$	$-p_y$	p_z	d_{xy}	$-d_{yz}$	$-d_{zx}$	$d_{x^2-y^2}$	d_{z^2}
σ_v	p_x	$-p_y$	p_z	$-d_{xy}$	$-d_{yz}$	d_{zx}	$d_{x^2-y^2}$	d_{z^2}
σ_v'	$-p_x$	p_y	p_z	$-d_{xy}$	d_{yz}	$-d_{zx}$	$d_{x^2-y^2}$	d_{z^2}

Table 2.6 Orbital Transformation (D_{4h})

Symmetry operation	p_x	p_y	p_z	d_{xy}	d_{yz}	d_{xz}	$d_{x^2-y^2}$	d_{z^2}
C_4	p_y	$-p_x$	p_z	$-d_{xy}$	d_{xz}	d_{yz}	$-d_{x^2-y^2}$	d_{z^2}
C_2	$-p_x$	$-p_y$	p_z	d_{xy}	$-d_{yz}$	$-d_{xz}$	$d_{x^2-y^2}$	d_{z^2}
C_2'	p_x	$-p_y$	$-p_z$	$-d_{xy}$	d_{yz}	$-d_{xz}$	$d_{x^2-y^2}$	d_{z^2}
C_2''	p_y	p_x	$-p_z$	d_{xy}	$-d_{xz}$	$-d_{yz}$	$-d_{x^2-y^2}$	d_{z^2}
i	$-p_x$	$-p_y$	$-p_z$	d_{xy}	d_{yz}	d_{xz}	$d_{x^2-y^2}$	d_{z^2}
S_4	$-p_y$	$-p_x$	$-p_z$	$-d_{xy}$	$-d_{xz}$	$-d_{yz}$	$-d_{x^2-y^2}$	d_{z^2}
σ_h	p_x	p_y	$-p_z$	d_{xy}	$-d_{yz}$	$-d_{xz}$	$d_{x^2-y^2}$	d_{z^2}
σ_v	p_x	$-p_y$	p_z	$-d_{xy}$	$-d_{yz}$	d_{xz}	$d_{x^2-y^2}$	d_{z^2}
σ_d	p_y	p_x	p_z	d_{xy}	d_{xz}	d_{yz}	$-d_{x^2-y^2}$	d_{z^2}

respectively. These tables, which should be verified by the reader, contain all the information necessary to enable us to write down the transformation equation (2.9) and deduce the character of the transformation matrix.

C_{2v}

Consider the p_y orbital. The operation E (the identity operation), as always, leaves it unaltered; therefore,

$$\hat{E}(\psi_{p_y}) = (1)\psi_{p_y} \tag{2.11}$$

Table 2.7   Orbital Transformation D_{3h}

Symmetry operation	p_x	p_y	p_z	d_{xy}
C_3	$-\dfrac{1}{2}p_x - \dfrac{\sqrt{3}}{2}p_y$	$\dfrac{\sqrt{3}}{2}p_x - \dfrac{1}{2}p_y$	p_z	$-\dfrac{1}{2}d_{xy} + \dfrac{\sqrt{3}}{2}d_{x^2-y^2}$
C_2	p_x	$-p_y$	$-p_z$	$-d_{xy}$
σ_h	p_x	p_y	$-p_z$	d_{xy}
S_3	$-\dfrac{1}{2}p_x - \dfrac{\sqrt{3}}{2}p_y$	$\dfrac{\sqrt{3}}{2}p_x - \dfrac{1}{2}p_y$	$-p_z$	$-\dfrac{1}{2}d_{xy} + \dfrac{\sqrt{3}}{2}d_{x^2-y^2}$
σ_v	p_x	$-p_y$	p_z	$-d_{xy}$

The operation $\hat{C}_2$ inverts the orbital so that the lobes are interchanged; therefore,

$$\hat{C}_2(\psi_{p_y}) = (-1)(\psi_{p_y}) \tag{2.12}$$

The operation $\hat{\sigma}_v$ has the same effect, whereas $\hat{\sigma}'_v$ leaves the orbital unaltered. The list of characters is therefore

$$\begin{array}{cccc} E & C_2 & \sigma_v & \sigma'_v \\ 1 & -1 & -1 & 1 \end{array} \tag{2.13}$$

which corresponds to the b_2 representation. The rest of the orbitals can be treated in exactly the same way. Table 2.8 gives the final result.

Table 2.8

Symmetry orbital (C_{2v})	Atomic orbital
a_1	$p_z, d_{z^2}, d_{x^2-y^2}$
a_2	d_{xy}
b_1	d_{xz}, p_x
b_2	d_{yz}, p_y

D_{4h}

The procedure for working out the characters in this point group is the same as before, but a new point arises in this case. Consider the p_x orbital. The operation $\hat{C}_4$ transforms it to the position originally occupied

Table 2.7 (*cont.*)

Symmetry operation	$d_{x^2-y^2}$	d_{xz}	d_{yz}	d_{z^2}
C_3	$-\dfrac{\sqrt{3}}{2}d_{xy} - \dfrac{1}{2}d_{x^2-y^2}$	$-\dfrac{1}{2}d_{xz} - \dfrac{\sqrt{3}}{2}d_{yz}$	$\dfrac{\sqrt{3}}{2}d_{xz} - \dfrac{1}{2}d_{yz}$	d_{z^2}
C_2	$d_{x^2-y^2}$	$-d_{xz}$	d_{yz}	d_{z^2}
σ_h	$d_{x^2-y^2}$	$-d_{xz}$	$-d_{yz}$	d_{z^2}
S_3	$-\dfrac{\sqrt{3}}{2}d_{xy} - \dfrac{1}{2}d_{x^2-y^2}$	$-\dfrac{1}{2}d_{xz} - \dfrac{\sqrt{3}}{2}d_{yz}$	$\dfrac{\sqrt{3}}{2}d_{xz} - \dfrac{1}{2}d_{yz}$	d_{z^2}
σ_v	$d_{x^2-y^2}$	d_{xz}	$-d_{yz}$	d_{z^2}

by the p_y orbital. Therefore, we can write

$$\hat{C}_4(\psi_{p_x}) = (1)\psi_{p_y} \qquad (2.14)$$

This indicates that the p_x and p_y orbitals are not independent, and the complete transformation equation is therefore

$$\hat{C}_4\begin{pmatrix} \psi_{p_x} \\ \psi_{p_y} \end{pmatrix} = \begin{pmatrix} 0 & 1 \\ -1 & 0 \end{pmatrix}\begin{pmatrix} \psi_{p_x} \\ \psi_{p_y} \end{pmatrix} \qquad (2.15)$$

The transformation matrix has a trace of zero.

Consider now one of the $\hat{C}_2'$ operations, that which lies along the x axis. This transforms p_x into itself and p_y into its inverse. The complete transformation equation is

$$\hat{C}_2'\begin{pmatrix} \psi_{p_x} \\ \psi_{p_y} \end{pmatrix} = \begin{pmatrix} 1 & 1 \\ 0 & -1 \end{pmatrix}\begin{pmatrix} \psi_{p_x} \\ \psi_{p_y} \end{pmatrix} \qquad (2.16)$$

and the character is zero. Either of the two $\hat{C}_2'$ operations could have been used to derive the transformation equation, but the one taken must be used consistently.

Finally, for this point group consider the operation $\hat{\imath}$. It transforms p_x to $-p_x$ and p_y to $-p_y$. Hence,

$$\hat{\imath}\begin{pmatrix} \psi_{p_x} \\ \psi_{p_y} \end{pmatrix} = \begin{pmatrix} -1 & 0 \\ 0 & -1 \end{pmatrix}\begin{pmatrix} \psi_{p_x} \\ \psi_{p_y} \end{pmatrix} \qquad (2.17)$$

The transformation matrix has a trace of -2. Table 2.9 lists the final results.

Table 2.9

Symmetry orbital (D_{4h})	Atomic orbital
a_{1g}	s, d_{z^2}
b_{1g}	$d_{x^2-y^2}$
b_{2g}	d_{xy}
e_g	(d_{xz}, d_{yz})
a_{2u}	p_z
e_u	(p_x, p_y)

D_{3h}

In the two previous cases, we have considered point groups in which the orbitals are transformed into positions which they themselves or other orbitals previously occupied. For the D_{3h} point group, the main rotation axis is C_3, a rotation by $2\pi/3$; it does not transform all p and d orbitals into recognizable positions. In this case, it is more convenient to use the wave functions expressed in polar coordinates, and it is only necessary to consider the angular part of the wave function. Consider first the p_z orbital, $\psi_{p_z} = R(r)\cos\theta$. The rotation $\hat{C}_3$ does not affect the angle θ; therefore, $\hat{C}_3(\psi_{p_z}) = (1)\psi_{p_z}$. For the p_x orbital, $\psi_{p_x} = R(r)\sin\theta\cos\phi$, the operation $\hat{C}_3$ increases ϕ by $2\pi/3$, and θ is unaltered. Hence, the point (r, θ, ϕ) is transformed to $(r, \theta, \phi + 2\pi/3)$, and the new wave function is

$$R(r)\sin\theta\cos\left(\phi + \frac{2\pi}{3}\right) = R(r)\sin\theta\left(\cos\phi\cos\frac{2\pi}{3} - \sin\phi\sin\frac{2\pi}{3}\right)$$

$$= R(r)\sin\theta\left(-\frac{1}{2}\cos\phi - \frac{\sqrt{3}}{2}\sin\phi\right)$$

$$= -\frac{1}{2}R(r)\sin\theta\cos\phi - \frac{\sqrt{3}}{2}R(r)\sin\theta\sin\phi$$

$$= -\frac{1}{2}\psi_{p_x} - \frac{\sqrt{3}}{2}\psi_{p_y} \qquad (2.18)$$

The p_y orbital can be treated in a similar way. The complete transformation equation is

$$\hat{C}_3\begin{pmatrix}\psi_{p_x}\\\psi_{p_y}\end{pmatrix} = \begin{vmatrix}-\dfrac{1}{2} & -\dfrac{\sqrt{3}}{2}\\[2mm]\dfrac{\sqrt{3}}{2} & -\dfrac{1}{2}\end{vmatrix}\begin{pmatrix}\psi_{p_x}\\\psi_{p_y}\end{pmatrix} \qquad (2.19)$$

and the character is -1. The transformations by the operators $\hat{C}_2$, $\hat{\sigma}_h$, and $\hat{\sigma}_v$ can be carried out by inspection, as described previously for the D_{4h} point group. The operator $\hat{S}_3$ is a rotation by $2\pi/3$ followed by reflection across the xy plane so that $\theta \rightarrow -\theta$ and $(\phi + 2\pi/3) \rightarrow -(\phi + 2\pi/3)$. Remembering that $\cos(-x) = \cos x$ and $\sin(-x) = -\sin x$, we can use the same procedure as before to determine the transformation matrix. The final results are listed in Table 2.10.

Table 2.10

Symmetry orbital (D_{3h})	Atomic orbital
a_1'	s, d_{z^2}
e'	$(d_{xy}, d_{x^2-y^2}), (p_x, p_y)$
a_2''	p_z
e''	(d_{xz}, d_{yz})

Table 2.11 lists the symmetry orbitals for the most common point groups. Any one of the sets can be deduced in the way described. Constant use will be made of this table in the following chapters.

Table 2.11 Symmetry Orbitals

Point group	s	p_z	p_x	p_y	d_{z^2}	$d_{x^2-y^2}$	d_{xy}	d_{yz}	d_{zx}
C_s	a'	a''	a'	a'	a'	a'	a'	a''	a''
C_i	a_g	a_u	a_u	a_u	a_g	a_g	a_g	a_g	a_g
C_2	a	a	b	b	a	a	a	b	b
C_3	a	a	e		a	e		e	
C_4	a	a	e		a	b	b	e	
C_5	a	a	e_1		a	e_2		e_1	
C_6	a	a	e_1		a	e_2		e_1	
D_2	a	b_1	b_3	b_2	a	a	b_1	b_3	b_2
D_3	a_1	a_2	e		a_1	e		e	
D_4	a_1	a_2	e		a_1	b_1	b_2	e	
D_5	a_1	a_2	e_1		a_1	e_2		e_1	
D_6	a_1	a_2	e_1		a_1	e_2		e_1	
C_{2v}	a_1	a_1	b_1	b_2	a_1	a_1	a_2	b_2	b_1
C_{3v}	a_1	a_1	e		a_1	e		e	
C_{4v}	a_1	a_1	e		a_1	b_1	b_2	e	
C_{5v}	a_1	a_1	e_1		a_1	e_2		e_1	
C_{6v}	a_1	a_1	e_1		a_1	e_2		e_1	
C_{2h}	a_g	a_u	b_u	b_u	a_g	a_g	a_g	b_g	b_g
C_{3h}	a'	a''	e'		a'	e'		e''	

Table 2.11 (*cont.*)

Point group	s	p_z	p_x	p_y	d_{z^2}	$d_{x^2-y^2}$	d_{xy}	d_{yz}	d_{zx}
C_{4h}	a_g	a_u	e_u		a_g	b_g	b_g	e_g	
C_{5h}	a'	a''	e'_1		a'	e'_2		e''_1	
C_{6h}	a_g	a_u	e_{1u}		a_g	e_{2g}		e_{1g}	
D_{2h}	a_g	b_{1u}	b_{3u}	b_{2u}	a_g	a_g	b_{1g}	b_{3g}	b_{2g}
D_{3h}	a'_1	a''_2	e'		a'_1	e'		e''	
D_{4h}	a_{1g}	a_{2u}	e_u		a_{1g}	b_{1g}	b_{2g}	e_g	
D_{5h}	a'_1	a''_2	e'_1		a'_1	e'_2		e''_1	
D_{6h}	a_{1g}	a_{2u}	e_{1u}		a_{1g}	e_{2g}		e_{1g}	
D_{2d}	a_1	b_2	e		a_1	b_1	b_2	e	
D_{3d}	a_{1g}	a_{2u}	e_u		a_{1g}	e_g		e_g	
D_{4d}	a_1	b_2	e_1		a_1	e_2		e_3	
D_{5d}	a_{1g}	a_{2u}	e_{2u}		a_{1g}	e_{2g}		e_{1g}	
D_{6d}	a_1	b_2	e_1		a_1	e_2		e_5	
S_4	a	b	e		a	b	b	e	
S_6	a_g	a_u	e_u		a_g	e_g		e_g	
T_d	a_1		t_2		e			t_2	
O_h	a_{1g}		t_{1u}			e_g		t_{2g}	
$C_{\infty v}$	a_1	a_1	e_1		a_1	e_2		e_1	
$D_{\infty h}$	Σ_g^+	Σ_u^+	Π_u		Σ_g^+	Δ_g		Π_g	
I_h	a_g		t_{1u}				h_g		

BIBLIOGRAPHY

1. H. EYRING, J. WALTER, and G. E. KIMBALL, *Quantum Chemistry*, John Wiley & Sons, Inc., New York, 1944.

2. J. N. MURRELL, S. F. A. KETTLE, and J. M. TEDDER, *Valence Theory*, 2nd ed., John Wiley & Sons, Inc., New York, 1970.

3. F. L. PILAR, *Elementary Quantum Chemistry*, McGraw-Hill Book Company, New York, 1968.

4. J. C. SLATER, *Quantum Theory of Atomic Structure*, Vol. 1, McGraw-Hill Book Company, New York, 1960.

5. F. A. COTTON, *Chemical Applications of Group Theory*, 2nd ed., Interscience Publishers, New York, 1971.

FURTHER READING

C. J. H. SCHUTTE, *The Wave Mechanics of Atoms, Molecules and Ions*, Arnold, Ltd., London, 1968.

L. D. LANDAU and E. M. LIFSHITZ, *Quantum Mechanics*, Pergamon Press, Inc., Elmsford, New York, 1965.

C. A. COULSON, *Valence*, Oxford University Press, Inc., London, 1961.

D. J. ROYER, *Bonding Theory*, McGraw-Hill Book Company, New York, 1968.

D. S. URCH, *Orbitals and Symmetry*, Penguin Books, Baltimore, Md., 1970.

3
Valence Bond Theory

3.1 INTRODUCTION

A molecule is an aggregate of two or more atoms attached in certain positions in space by chemical bonds. Bond types range from the attraction of ionic species to nonpolar sharing of electrons. It is not possible to draw a sharp distinction between these two extremes for the majority of chemical compounds. The covalent bond, our main concern, is normally represented as an overlap of atomic orbitals. In this way, electrons that were initially attached to one atom are now shared by two or more atoms.

Since atomic orbitals have directional properties it is necessary to know in what way orbitals overlap in a bond, as this will give us a clue to the final shape of a molecule. A reasonable assumption to start from is that the bond formed is the strongest possible. Therefore, the requirements for the formation of a strong bond are important to our inquiry.

In Chapters 3 through 6, the basic concepts used to describe the covalent bond will be briefly outlined, with particular emphasis on stereochemical aspects. This chapter will deal with the valence bond (VB) theory, in which the bonding electrons are considered to be localized between two bonding centers.

3.2 DIATOMIC MOLECULE

The Schrödinger equation for a diatomic molecule AB expressed in the Hamiltonian form is

$$\hat{H}\psi_{AB} = E\psi_{AB} \tag{3.1}$$

where

$$\hat{H} = -\frac{h^2}{8\pi^2 m}(\nabla_1^2 + \nabla_2^2) - \frac{e^2}{r_{A1}} - \frac{e^2}{r_{B2}}$$

$$-e^2\left(\frac{1}{r_{A2}} + \frac{1}{r_{B1}} - \frac{1}{r_{AB}} - \frac{1}{r_{12}}\right) \tag{3.2}$$

and the symbols refer to Fig. 3.1. The molecular wave function in the valence bond theory is

$$\psi_{AB} = \psi_A(1)\psi_B(2) \pm \psi_A(2)\psi_B(1) \tag{3.3}$$

where (1) and (2) refer to the bonding electrons and ψ_A and ψ_B to the atomic orbitals involved in the bond. Other terms that could contribute to ψ_{AB} are $\psi_A(1)\psi_A(2)$ and $\psi_B(1)\psi_B(2)$, corresponding to A^-B^+ and A^+B^-, respectively.

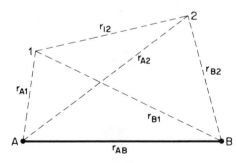

Fig. 3.1 Coordinates for molecule AB with two electrons 1 and 2.

Solution of Equation (3.1)[1] for E gives

$$E^{\pm} = E_0^A + E_0^B + \frac{Q \pm A}{1 \pm S^2} \tag{3.4}$$

where E_0^A and E_0^B are the ground state energies for the atoms A and B, respectively. The other terms in Equation (3.4) are the Coulomb integral

$$Q = \iint \psi_A(1)\psi_B(2)|\hat{H}'|\psi_A(1)\psi_B(2)\, dT_1\, dT_2$$

the exchange integral

$$A = \iint \psi_A(1)\psi_B(2)|\hat{H}'|\psi_A(2)\psi_B(1)\, dT_1\, dT_2$$

and the overlap integral

$$S = \int \psi_A(1)\psi_B(1)\, dT_1 = \int \psi_A(2)\psi_B(2)\, dT_2$$

where

$$\hat{H}' = \hat{H} - E_0^A - E_0^B$$

We can deduce a number of useful facts from a qualitative study of Equation (3.4). The formation of a bond by the overlap of two atomic orbitals gives two molecular orbitals, one bonding and one antibonding (or nonbonding) with energies E^+ and E^-, respectively. The term that is mainly responsible for the difference in the two energies is the exchange energy A. The exchange energy is an important factor in determining bond stability. The plots of both Coulomb energy and total binding energy against the internuclear distance (Fig. 3.2) indicate this.

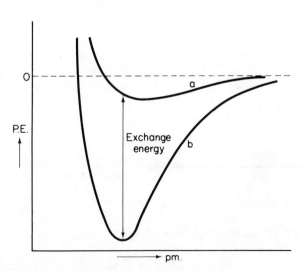

Fig. 3.2 Coulomb energy (a) and total binding energy (b) plotted against internuclear distance.

The exchange energy term can be expanded as follows:

$$A = \int\int \psi_A(1)\psi_B(2)|\hat{H}'|\psi_A(2)\psi_B(1)\, dT_1\, dT_2$$

$$= \int\int \psi_A(1)\psi_B(2)\left|e^2\left(\frac{1}{r_{AB}} + \frac{1}{r_{12}} - \frac{1}{r_{A2}} - \frac{1}{r_{B1}}\right)\right|\psi_A(2)\psi_B(1)\, dT_1\, dT_2$$

$$= \frac{e^2 S^2}{r_{AB}} - S\int \frac{e^2}{r_{A2}}\psi_A(2)\psi_B(2)\, dT_2$$

$$\quad - S\int \frac{e^2}{r_{B1}}\psi_A(1)\psi_B(1)\, dT_1$$

$$\quad + \int\int \frac{e^2}{r_{12}}\psi_A(1)\psi_B(2)\psi_A(2)\psi_B(1)\, dT_1\, dT_2 \tag{3.5}$$

Equation (3.5) shows that the exchange energy is approximately proportional to the overlap integral. Therefore, the greater is the overlap between the atomic orbitals the larger is the exchange energy and hence the total binding energy. This is an important result and is basic to the VB approach to stereochemistry.

The criterion of maximum overlap is often used as a guide to bond strength and stereochemistry. For the overlap to be a maximum, the internuclear distance should be small, the bonding orbitals should be comparable in size and energy, and the orbitals must have the same symmetry.

3.3 POLYATOMIC MOLECULES

The VB method can be applied to polyatomic molecules by making use of the concept of localized bonding orbitals. For example, the three N—H bonds of NH_3 are considered to be independent of each other. Evidence in support of this is that the N—H bond energy is approximately one-third the enthalpy of atomization of the ammonia molecule.

The nitrogen atom in its ground state electronic configuration $(1s)^2(2s)^2(2p)^3$ has three unpaired electrons, one in each of the p orbitals. Three N—H bonds can be formed using these p orbitals and the $1s$ orbitals of the three hydrogens. The wave functions will have the form

$$\psi_{\mathrm{I}} = a\psi_{2p_xN}(1)\psi_{1sH}(2) \pm b\psi_{2p_xN}(2)\psi_{1sH}(1)$$

$$\psi_{\mathrm{II}} = a\psi_{2p_yN}(3)\psi_{1sH}(4) \pm b\psi_{2p_yN}(4)\psi_{1sH}(3) \tag{3.6}$$

$$\psi_{\mathrm{III}} = a\psi_{2p_zN}(5)\psi_{1sH}(6) \pm b\psi_{2p_zN}(6)\psi_{1sH}(5)$$

The energy of the bonding molecular orbital ψ_I^+ is given by an equation similar to (3.4), viz.,

$$E_I^+ = E_N + E_H + \frac{Q + A}{1 + S^2} \tag{3.7}$$

This energy may be related to the N—H bond energy. The bond will be strongest when the overlap is greatest, which is achieved when the hydrogen atoms approach along the x, y, and z axes. Hence, the HNH bond angle is predicted to be 90°, the angle between p orbitals. The actual angle is 106.6°. A similar argument predicts the HOH bond angle in H_2O to be 90°, when, in fact, it is 104.5°. The deviations indicate that in its present state, the theory is not completely adequate, although it does predict the gross stereochemistry.

3.4   PERFECT PAIRING[7]

If for a molecule the energies of the free atoms are ignored as being independent of the molecular configuration, and if the overlap integral is neglected as being small relative to the Coulomb and exchange energies, then Equation (3.4) becomes

$$E = Q + \sum_b A_{ij}^b - \tfrac{1}{2} \sum_{nb} A_{ij}^{nb} \tag{3.8}$$

where A_{ij}^b represents all the exchange terms between bonded atoms i and j and A_{ij}^{nb} signifies all the exchange terms between nonbonded atoms i and j. It is also possible to neglect Q, for it is only about 10% of the exchange energy. Thus, Equation (3.8) reduces to

$$E = \sum_b A_{ij}^b - \tfrac{1}{2} \sum_{nb} A_{ij}^{nb} \tag{3.9}$$

This approximation is not too extreme when we compare two molecular configurations that differ only in bond angle. Consider a bond between two atoms X and Y such that X uses a $1s$ orbital and Y a $2p$ orbital. If the half-filled p orbital of Y is p_x, then the greatest overlap occurs when X approaches along the x axis. However, place X at an angle α to the x axis, as shown in Fig. 3.3. The exchange energy given in Equation (3.5) is ($S = 0$)

$$A = \int\int \psi_{1s}(1)\psi_{p_x}(2)\left|\frac{e^2}{r_{12}}\right|\psi_{1s}(2)\psi_{p_x}(1)\,dT_1\,dT_2 \tag{3.10}$$

Resolve the p_x orbital into two components, p' lying along the XY direction and p'' at right angles to it. That is,

$$\psi_{p_x} = \psi_{p'}\cos\alpha + \psi_{p''}\sin\alpha \tag{3.11}$$

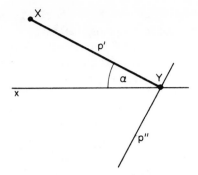

Fig. 3.3 Position of *X* with respect to *x* axis.

Place (3.11) into Equation (3.10) to give

$$A(\alpha) = \int\int \psi_{1s}(1)[\psi_{p'}(2)\cos\alpha + \psi_{p''}(2)\sin\alpha]\left|\frac{e^2}{r_{12}}\right|\psi_{1s}(2)$$

$$\times\ (\psi_{p'}(1)\cos\alpha + \psi_{p''}(1)\sin\alpha)\,dT_1\,dT_2 \qquad (3.12)$$

Expand Equation (3.12)

$$A(\alpha) = \int\int \psi_{1s}(1)\psi_{p'}(2)\psi_{1s}(2)\psi_{p'}(1)\frac{\cos^2\alpha e^2}{r_{12}}\,dT_1\,dT_2$$

$$+ \int\int \psi_{1s}(1)\psi_{p''}(2)\psi_{1s}(2)\psi_{p''}(1)\frac{\sin^2\alpha e^2}{r_{12}}\,dT_1\,dT_2 \qquad (3.13)$$

Terms in sin α cos α vanish and Equation (3.13) simplifies to

$$A(\alpha) = A(0°)\cos^2\alpha + A(90°)\sin^2\alpha \qquad (3.14)$$

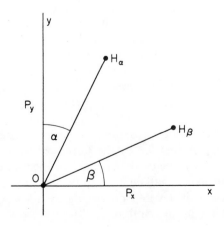

Fig. 3.4 Position of H_α and H_β for perfect pairing approach.

The exchange energy $A(\alpha)$ is a maximum when $\alpha = 0°$—i.e., when X lies along the x axis. This approach is called the perfect pairing method.

When applied to water, the angles α and β between the lines joining the oxygen and protons H_α and H_β and x and y axes, respectively (Fig. 3.4), are zero for maximum exchange energy. Therefore, the HOH bond angle is again predicted to be 90°.

3.5   HYBRIDIZATION

The valence bond theory, as described so far, has a difficulty in handling the bonding in molecules like CH_4. The elementary approach would be to write down four bonding orbitals as follows:

$$\psi_I = \psi_{C_1}(1)\psi_{H_{1s}}(2) + \psi_{C_1}(2)\psi_{H_{1s}}(1)$$
$$\psi_{II} = \psi_{C_2}(3)\psi_{H_{1s}}(4) + \psi_{C_2}(4)\psi_{H_{1s}}(3)$$
$$\psi_{III} = \psi_{C_3}(5)\psi_{H_{1s}}(6) + \psi_{C_3}(6)\psi_{H_{1s}}(5)$$
$$\psi_{IV} = \psi_{C_4}(7)\psi_{H_{1s}}(8) + \psi_{C_4}(8)\psi_{H_{1s}}(7)$$

(3.15)

where $\psi_{H_{1s}}$ is the $1s$ orbital of each hydrogen and ψ_{C_1}, ψ_{C_2}, ψ_{C_3}, ψ_{C_4} are the four carbon orbitals. The ground state of carbon is $(1s)^2(2s)^2(2p)^1(2p)^1$, which does not have the correct number of singly occupied orbitals. In fact, carbon would be expected to be divalent. To achieve quadrivalence, it is necessary to promote one of the $2s$ electrons to the vacant $2p$ orbital, which requires approximately $405\,kJ\,mol^{-1}$.

Use of Equations (3.15) now would lead to one C—H bond being different from the other three, since ψ_{C_1} is a $2s$ orbital and the rest are $2p$ orbitals. The stereochemistry on this basis would be three bonds at right angles and the fourth at a 125° angle to the rest. This is contrary to experiment, for all the bonds are of equal strength and are at 109° 28′ to each other. It is at this point that hybridization becomes a necessary step in the VB theory. The $2s$ and three $2p$ orbitals are combined in a linear combination to give a set of four equivalent orbitals called sp^3 hybrids. The change in shape and directional character in going from a p orbital to a sp^3 hybrid is depicted in Fig. 3.5. The energy required to convert the excited ^5S state to the hypothetical hybridized state (called the "valence state") is approximately $270\,kJ\,mol^{-1}$. However, this is compensated for by the improved overlap with the hybrid orbitals. It is necessary to be clear that the "valence state" has no real existence and hybridization is a mathematical process which is a convenient way of describing stereochemistry and preserving the idea of a localized bond and the criterion of maximum overlap.

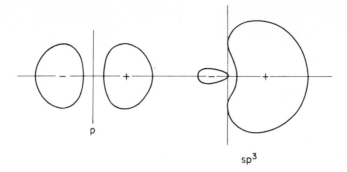

Fig. 3.5 Angular functions for a 2*p* orbital and a *sp*³ hybrid orbital.

The method of perfect pairing or maximum overlap can now be used to determine the stereochemistry using the hybrid orbitals in exactly the same way as described previously.

3.6 CONSTRUCTION OF HYBRID ORBITALS

It is first necessary to determine which of the atomic orbitals can be used for a particular set of hybrid orbitals. This is readily achieved with the use of group theory.[8] First, a reducible representation is obtained by considering how the bonds in a molecule transform under the operations of the point group of the molecule. The representation (reducible) is then reduced to its components, and by making use of Table 2.11 we can find the atomic orbitals with the same transformation properties.

Consider as an example a tetrahedral molecule (Fig. 3.6). The four bonds are unchanged by the identity operation $\hat{E}$. Hence, we can write

$$\hat{E}\begin{pmatrix}1\\2\\3\\4\end{pmatrix}=\begin{pmatrix}1&0&0&0\\0&1&0&0\\0&0&1&0\\0&0&0&1\end{pmatrix}\begin{pmatrix}1\\2\\3\\4\end{pmatrix}\qquad \chi=4$$

The operation $\hat{C}_3$ (shown in Fig. 3.6) changes three bonds and leaves one unaltered. The transformation equation is

$$\hat{C}_3\begin{pmatrix}1\\2\\3\\4\end{pmatrix}=\begin{pmatrix}1&0&0&0\\0&0&1&0\\0&0&0&1\\0&1&0&0\end{pmatrix}\begin{pmatrix}1\\2\\3\\4\end{pmatrix}\qquad \chi=1$$

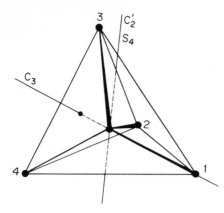

Fig. 3.6 Symmetry axes for a tetrahedral molecule.

It is not necessary to write the transformation equations in full, as χ = number of bonds unchanged.

Hence, for a tetrahedral molecule we have the reducible representation

$$E \quad 8C_3 \quad 3C_2 \quad 6S_4 \quad 6\sigma_d$$

$$\Gamma_{tet} \quad 4 \quad 1 \quad 0 \quad 0 \quad 2$$

which reduces to $A_1 + T_2$. From Table 2.11, the atomic orbitals that clearly can form the tetrahedral hybrids are

a_1	t_2
s	$p_x\, p_y\, p_z$
	$d_{xy}\, d_{yz}\, d_{xz}$

The hybrid orbitals will be sp^3 or sd^3, and, in general, there is no problem in deciding which of the two possible hybrids an atom may use. For example, because of the large energy difference between the $2p$ and $3d$ orbitals of carbon (approximately 960 kJ mol^{-1}), the hybrids will be exclusively sp^3. It is possible, however, that if the p and d orbitals are of comparable energy, a mixture of both sp^3 and sd^3 hybrids can be considered. For example, the bonding $3d$, $4s$, and $4p$ orbitals of Cr(VI) and Mn(VII) in the anions CrO_4^{2-} and MnO_4^- are of comparable energy, so the tetrahedral hybrids are best represented by $sp^n d^m$.

Table 3.1 lists all the possible hybrid orbital combinations for co-ordination numbers 2 to 9.

Some of the various methods of obtaining the mathematical form of the sp^2 and sp^3 hybrid orbitals will now be outlined.

Table 3.1[9]   Hybrid Orbitals

Coordination number	Point group	σ-Hybrid orbitals	Stereochemistry	π-Hybrid orbitals ($\perp$)	($//$)
2	$D_{\infty h}$	sp	Linear	pd	pd
	$D_{\infty h}$	dp	Linear	pd	pd
	C_{2v}	p^2	Angular	pd^2	d^2
	C_{2v}	ds	Angular	p^2d	pd^2
	C_{2v}	d^2	Angular	p^2d	pd
3	D_{3h}	sp^2	Trigonal plane	pd^2	d^2
	D_{3h}	dp^2	Trigonal plane	pd^2	d
	D_{3h}	d^2s	Trigonal plane	pd^2	p^2
	D_{3h}	d^3	Trigonal plane	pd^2	p^2
	C_{2v}	dsp	Unsymmetrical plane	pd^2	pd
	C_{3v}	p^3	Trigonal pyramid	d^5	
	C_{3v}	d^2p	Trigonal pyramid	p^2d^3	
4	T_d	sp^3	Tetrahedral	d^5	
	T_d	d^3s	Tetrahedral	p^3d^2	
	D_{4h}	dsp^2	Tetragonal plane	pd^2	d
	D_{4h}	d^2p^2	Tetragonal plane	dp^2	
	C_{3v}	d^2sp	Irregular tetrahedron	d	
	C_{3v}	dp^3	Irregular tetrahedron		
	C_{3v}	d^3p	Irregular tetrahedron		
	C_{4v}	d^4	Tetragonal pyramid	dp^2	
5	D_{3h}	dsp^3	Bipyramid	d^2	d^2
	D_{3h}	d^3sp	Bipyramid	d^2	p^2
	C_{4v}	d^2sp^2	Tetragonal pyramid	d	pd^2
	C_{4v}	d^4s	Tetragonal pyramid	d	p^3
	C_{4v}	d^2p^3	Tetragonal pyramid	d	d^2
	C_{4v}	d^4p	Tetragonal pyramid	d	p^2
	D_{5h}	d^3p^2	Pentagonal plane	pd^2	
	D_{5h}	d^5	Pentagonal pyramid	p^3	
6	O_h	d^2sp^3	Octahedron	d^3	
	D_{3h}	d^4sp	Trigonal prism		p^2d
	D_{3h}	d^5p	Trigonal prism		p^2
	D_{3d}	d^3p^3	Trigonal antiprism		d
		d^3sp^2	Mixed		
		d^5s	Mixed		
		d^4p^2	Mixed		
7	C_{3v}	d^3sp^3	ZrF_7^{3-} Structure		d^2
	C_{3v}	d^5sp	ZrF_7^{3-} Structure		p^2
	C_{2v}	d^4sp^2	TaF_7^{2-} Structure		dp
	C_{2v}	d^4p^3	TaF_7^{2-} Structure		ds
	C_{2v}	d^5p^2	TaF_7^{2-} Structure		ps
8	V_d	d^4sp^3	Dodecahedron	d	
	D_{4d}	d^5p^3	Antiprism		
	C_{3v}	d^5sp^2	Face-centered prism	p	
9		d^5sp^3	Tripyramid		

49

sp² Hybrids

The *sp²* hybrids have D_{3h} symmetry (Fig. 3.7). In general terms, the three hybrids will be of the form

$$\psi_1 = a_1 s + b_1 p_x + c_1 p_y$$
$$\psi_2 = a_2 s + b_2 p_x + c_2 p_y \qquad (3.16)$$
$$\psi_3 = a_3 s + b_3 p_x + c_3 p_y$$

Since ψ_1 is placed along the x axis, the p_y component in that direction is zero; hence, $c_1 = 0$. We now can make use of symmetry to help find the value of the

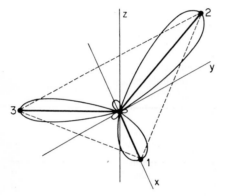

Fig. 3.7 Coordinate system for *sp²* hybrid orbitals.

other coefficients. For example, both ψ_2 and ψ_3 should transform into ψ_1 by any of the operations of the D_{3h} point group. Consider $\hat{C}_3$ operating on ψ_2. The appropriate equation is $\hat{C}_3(\psi_2) = \psi_1$; i.e.,

$$a_2[\hat{C}_3(s)] + b_2[\hat{C}_3(p_x)] + c_2[\hat{C}_3(p_y)] = a_1 s + b_1 p_x + 0 p_y \qquad (3.17)$$

Now

$$\hat{C}_3(s) = s$$

$$\hat{C}_3(p_x) = p_x \cos 120° - p_y \sin 120° = -\frac{1}{2}p_x - \frac{\sqrt{3}}{2}p_y$$

$$\hat{C}_3(p_y) = p_x \sin 120° + p_y \cos 120° = \frac{\sqrt{3}}{2}p_x - \frac{1}{2}p_y$$

Substituting into Equation (3.17) gives

$$a_2 s - b_2\left(\frac{1}{2}p_x + \frac{\sqrt{3}}{2}p_y\right) + c_2\left(\frac{\sqrt{3}}{2}p_x - \frac{1}{2}p_y\right) = a_1 s + b_1 p_x + 0 p_y \qquad (3.18)$$

Collect terms in s, p_x, and p_y on the left-hand side

$$a_2 s + p_x\left(-\frac{1}{2}b_2 + \frac{\sqrt{3}}{2}c_2\right) + p_y\left(-\frac{\sqrt{3}}{2}b_2 - \frac{1}{2}c_2\right) = a_1 s + b_1 p_x + 0 p_y$$

The coefficients of the same atomic orbital can be equated and arranged to give

$$a_1 = a_2$$
$$c_2 = -\sqrt{3}b_2 \qquad\qquad (3.19)$$
$$b_1 = -2b_2$$

For the transformation of ψ_3, we shall use the operation $\hat{\sigma}_{xz}$, reflection in the xz plane.

$$a_3[\hat{\sigma}_{xz}(s)] + b_3[\hat{\sigma}_{xz}(p_x)] + c_3[\hat{\sigma}_{xz}(p_y)] = a_2 s + b_2 p_x + c_2 p_y$$

Now

$$\hat{\sigma}_{xz}(s) = s$$
$$\hat{\sigma}_{xz}(p_x) = p_x$$
$$\hat{\sigma}_{xz}(p_y) = -p_y$$

Hence,

$$a_3 s + b_3 p_x - c_3 p_y = a_2 s + b_2 p_x + c_2 p_y$$

That is,

$$a_3 = a_2$$
$$b_3 = b_2 \qquad\qquad (3.20)$$
$$-c_3 = c_2$$

We require two further equations to solve for the eight coefficients. To obtain these, we make use of the normalization and orthogonality requirements. The normalization of ψ_2 requires

$$a_2^2 + b_2^2 + c_2^2 = 1 \qquad\qquad (3.21)$$

and the orthogonality of ψ_2 and ψ_3 requires

$$a_2 a_3 + b_2 b_3 + c_2 c_3 = 0 \qquad\qquad (3.22)$$

Solution of the eight equations (3.19), (3.20), (3.21), and (3.22) gives

$$a_1 = a_2 = a_3 = \frac{1}{\sqrt{3}}$$

$$b_1 = \frac{2}{\sqrt{6}}$$

$$b_2 = b_3 = -\frac{1}{\sqrt{6}}$$

$$c_2 = \frac{1}{\sqrt{2}}$$

$$c_3 = -\frac{1}{\sqrt{2}}$$

Therefore, the three hybrid orbitals are

$$\psi_1 = \frac{1}{\sqrt{3}}s + \frac{2}{\sqrt{6}}p_x$$

$$\psi_2 = \frac{1}{\sqrt{3}}s - \frac{1}{\sqrt{6}}p_x + \frac{1}{\sqrt{2}}p_y \qquad (3.23)$$

$$\psi_3 = \frac{1}{\sqrt{3}}s - \frac{1}{\sqrt{6}}p_x - \frac{1}{\sqrt{2}}p_y$$

The correctness of the Equations (3.23) can be checked by showing that $a^2 + b^2 + c^2 = 1$ for each orbital, and $a_1^2 + a_2^2 + a_3^2 = 1$ and similarly for coefficients b and c.

sp^3 Hybrids

The tetrahedral hybrids sp^3 may be obtained by inspection. Consider the disposition of the four hybrid orbitals with respect to the x, y, and z axes (Fig. 3.8) (the sign convention for the axes is indicated). The four orbitals can be expressed in the form

$$\psi_i = a_i s + b_i p_x + c_i p_y + d_i p_z \qquad (i = 1, 2, 3, 4)$$

The signs of the coefficients can be determined by inspection. They are positive when the positive lobe of the p orbitals lies closest to the hybrid that they refer to, and negative when the negative lobe lies nearest.

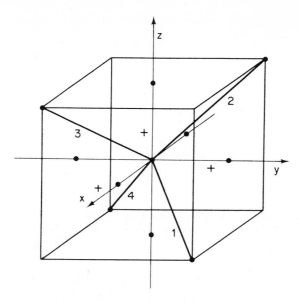

Fig. 3.8   Coordinate system for sp^3 hybrid orbitals.

Therefore, the signs are

$$
\begin{array}{ccccc}
\psi_1 & + & + & + & - \\
\psi_2 & + & - & + & + \\
\psi_3 & + & + & - & + \\
\psi_4 & + & - & - & - \\
\end{array}
$$

Since the p orbitals make equal contribution to each hybrid, the normalization constant is $\frac{1}{2}$. Hence,

$$
\psi_1 = \tfrac{1}{2}(s + p_x + p_y - p_z)
$$
$$
\psi_2 = \tfrac{1}{2}(s - p_x + p_y + p_z)
$$
$$
\psi_3 = \tfrac{1}{2}(s + p_x - p_y + p_z)
$$
$$
\psi_4 = \tfrac{1}{2}(s - p_x - p_y - p_z)
$$

The mathematical form of the hybrids depends on the position of the hybrids with respect to the Cartesian coordinates. For example, if ψ_1, for tetrahedral hybrids, is taken as lying along the x axis, the four hybrids are given by the expressions

$$\psi_1 = \frac{1}{2}s + \frac{\sqrt{3}}{2}p_x$$

$$\psi_2 = \frac{1}{2}s - \frac{1}{2\sqrt{3}}p_x + \sqrt{\frac{2}{3}}p_z$$

$$\psi_3 = \frac{1}{2}s - \frac{1}{2\sqrt{3}}p_x + \frac{1}{\sqrt{2}}p_y - \frac{1}{\sqrt{6}}p_z$$

$$\psi_4 = \frac{1}{2}s - \frac{1}{2\sqrt{3}}p_x - \frac{1}{\sqrt{2}}p_y - \frac{1}{\sqrt{6}}p_z$$

The reader can check that the functions are normalized and orthogonal.

The concept of hybrid orbitals can now be used to explain the stereochemistry of a great number of compounds. For example, boron trifluoride has a trigonal planar structure. The electronic configuration of boron is $(1s)^2(2s)^2(2p)^1$, and therefore the stereochemistry can be adequately explained by the use of sp^2 hybrids. Sulfur is six coordinate and octahedral in SF_6, and it is necessary to make use of the diffuse outer $3d$ orbitals of sulfur for the construction of six hybrids. The two $3d$ orbitals involved are $3d_{x^2-y^2}$ and $3d_{z^2}$. It is suggested that the "effective" positive charge on the sulfur atom is large in SF_6 owing to the high electronegativity of the fluorine atoms. The high positive charge will cause a contraction of the $3d$ orbitals to a more suitable size. However, an uninhibited use of outer orbitals in hybrid schemes can be dangerous, because the requirements for a good set of hybrid orbitals are similar to those given in Section 3.2 for good overlap of orbitals.

The stereochemistry of water can also be considered in terms of hybrid orbitals. The $2s$ and $2p$ orbitals of oxygen can be considered to form tetrahedral hybrids, two of which bond the protons and two of which accommodate the two lone pairs. The expected bond angle 109° 28′ is not too different from the experimental value of 104.5°. The decrease may be explained by the repulsion between the lone pairs that pushes the protons together. However, there is some doubt as to whether the energy required to produce oxygen in the tetrahedral valence state is compensated for by the formation of only two O—H bonds.

3.7 HYBRID ORBITALS AND OVERLAP

The hybrid orbitals given in Table 3.1 are composed of integral numbers of atomic orbitals; however, this is not necessary. The common method used to determine the composition of hybrid orbitals is to maximize the overlap.

For example, the best set of square planar hybrids have been determined as $s^{4/9}d^{14/9}p^2$, whereas for tetrahedral hybrids using both p and d orbitals the set giving maximum bond strength has been found to be $sd^{15/8}p^{9/8}$. The arrangement of six bonds can be octahedral sp^3d^2, trigonal prismatic $s^np^{1+2m}d^{5-n-2m}$, where $m = 0.5972$ and $n = 0.8464$, or trigonal antiprismatic $s^np^3d^{3-n}$, where $n = 0.8427$. The last set of hybrids is only slightly better than the octahedral hybrids at forming strong bonds. But both are better than the trigonal prismatic arrangement, and therefore are the more likely stereochemistries for six coordination.[10]

Until now we have been considering a set of hybrid orbitals as equivalent, which is not always the case. For example, in an AB_5 molecule having a trigonal bipyramid structure, the axial bonds are normally longer than the equatorial ones. Clearly, in this case the hybrids cannot be equivalent, and two sets can be employed, one for the trigonal plane and one for the linear bonds—i.e., $sp^2 + p_z d_{z^2}$ or $d_{z^2}p^2 + sp_z$, rather than the equivalent set $sp^3d_{z^2}$. Since the axial bonds are longer than the planar ones, the maximization of overlap may be used to suggest the more probable of the two sets of hybrids.

It was mentioned earlier that the stereochemistry of water could be discussed in terms of tetrahedral hybrids. Since there are two types of orbitals, bonding and nonbonding, a more accurate picture would be to determine two hybrid sets, one to bond the protons and one to hold the lone pairs. If this is done, the bonding hybrids would have less s character than the tetrahedral hybrids and the lone pair hybrids would have less p character.

3.8 π BONDS

Hybrid orbitals can also be formed for π-type atomic orbitals. Consider a square planar species with D_{4h} symmetry, such as $PtCl_4^{2-}$. The π bonds possible in the structure are represented as vectors in Fig. 3.9. The transformation properties of the π bonds are found by the methods used for σ bonds on p. 47. The reducible representations $\Gamma_{\pi(\perp)}$ and $\Gamma_{\pi(//)}$ are

	E	$2C_4$	C_2	$2C_2'$	$2C_2''$	i	$2S_4$	σ_h	$2\sigma_v$	$2\sigma_d$
$\Gamma_{\pi(\perp)}$	4	0	0	-2	0	0	0	-4	2	0
$\Gamma_{\pi(//)}$	4	0	0	-2	0	0	0	4	-2	0

which reduce to

$$\Gamma_{\pi(\perp)} = E_g + A_{2u} + B_{2u} \qquad \Gamma_{\pi(//)} = E_u + A_{2g} + B_{2g}$$

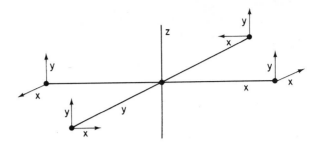

Fig. 3.9 Vectors representing π-orbitals for a square planar molecule.

From Table 2.11, we see that the atomic orbitals on the central atom corresponding to these irreducible representations are

$$\Gamma_{\pi(\perp)} \quad d_{xz}, d_{yz} \quad (e_g)$$
$$p_z \quad (a_{2u})$$
$$\Gamma_{\pi(//)} \quad p_x, p_y \quad (e_u)$$
$$d_{xy} \quad (b_{2g})$$

The π bonds perpendicular to the plane of the molecule can be pd^2 hybrids, whereas in the plane of the molecule they can be p^2d. It is unlikely that all the π bonds can be formed, because some of the preceding orbitals will have been used in the σ bonding (e.g., p_x and p_y) or will be energetically inaccessible.

In Table 3.1, the π-hybrid orbitals are listed for the atomic orbitals not involved in σ bonding.

3.9 *f* ORBITALS

In a few cases, the bonding within compounds and the formation of hybrid orbitals have been discussed in terms of some f-orbital contribution. The inclusion of f orbitals into hybrids concentrates the charge cloud more effectively along the bond direction. For example, a tetrahedral hybrid composed of $sp^3 + 4\% d + 2\% f$ is more strongly directed along the internuclear axis and gives greater bond overlap. But this is achieved at the expense of incorporating higher energy orbitals. It is doubtful that the extra energy required justifies the addition of the f component.

BIBLIOGRAPHY

1. H. EYRING, J. WALTER, and G. E. KIMBALL, *Quantum Chemistry*, John Wiley & Sons, Inc., New York, 1944.

2. J. N. MURRELL, *J. Chem. Phys.*, **32**, 767 (1960).

3. T. L. GILBERT and P. C. LYKOS, *J. Chem. Phys.*, **34**, 2199 (1961).

4. A. GOLEBIEWSKI, *Trans. Faraday Soc.*, **57**, 1849 (1961).

5. H. KUHN, *J. Chem. Phys.*, **16**, 727 (1948).

6. A. MACCOLL, *Trans. Faraday Soc.*, **46**, 369 (1950).

7. C. A. COULSON, *Valence*, 2nd ed., Oxford University Press, London, 1961.

8. F. A. COTTON, *Chemical Application of Group Theory*, 2nd ed., Interscience Publishers, New York, 1971.

9. G. E. KIMBALL, *J. Chem. Phys.*, **8**, 188 (1940).

10. G. H. DUFFEY, *J. Chem. Phys.*, **17**, 1328 (1949); **18**, 128, 510 (1950); **19**, 92 (1951).

FURTHER READING (In addition to list at end of Chapter 2)

H. H. JAFFE and M. ORCHIN, *Symmetry in Chemistry*, John Wiley & Sons, Inc., New York, 1965.

J. C. SLATER, *Quantum Theory of Molecules and Solids*, Vol. 1, McGraw-Hill Book Company, New York, 1963.

4

Pauli Exclusion Principle

4.1 PAULI EXCLUSION PRINCIPLE

A fundamental postulate in quantum chemistry that has a number of implications is the Pauli exclusion principle. The principle in its simplest form states that each electron has a unique set of quantum numbers, n, l, m, and s. Since each space orbital is described by the numbers n, l, and m, the principle tells us that the maximum number of electrons in an orbital is two, because there are only two values of the spin quantum number. A more general statement of the principle is that the complete wave function of a system $\psi = \psi$ space ψ spin, is antisymmetric. This means that when the coordinates of the two electrons in an orbital are interchanged, the wave function changes sign.

It is not obvious that these two statements of the Pauli exclusion principle say the same thing. In fact, the first statement is implied by the more general second one. The wave function for the singlet ground state of the H_2 molecule is

$$\psi = [\psi_A(1)\psi_B(2) + \psi_A(2)\psi_B(1)][\alpha(1)\beta(2) - \alpha(2)\beta(1)] \qquad (4.1)$$

where α and β represent the two spin states of an electron.

Expansion gives

$$\psi = \psi_A(1)\psi_B(2)\alpha(1)\beta(2) - \psi_A(1)\psi_B(2)\alpha(2)\beta(1)$$
$$+ \psi_A(2)\psi_B(1)\alpha(1)\beta(2) - \psi_A(2)\psi_B(1)\alpha(2)\beta(1) \tag{4.2}$$

which can be expressed in determinant form

$$\psi = \begin{vmatrix} \psi_A(1)\alpha(1) & \psi_B(1)\beta(1) \\ \psi_A(2)\alpha(2) & \psi_B(2)\beta(2) \end{vmatrix} - \begin{vmatrix} \psi_A(1)\beta(1) & \psi_B(1)\alpha(1) \\ \psi_A(2)\beta(2) & \psi_B(2)\alpha(2) \end{vmatrix} \tag{4.3}$$

If the three quantum numbers n, l, and m are identical for ψ_A and ψ_B, i.e., $\psi_A = \psi_B$, then Equation (4.3) becomes

$$\psi = \begin{vmatrix} \psi_A(1)\alpha(1) & \psi_A(1)\beta(1) \\ \psi_A(2)\alpha(2) & \psi_A(2)\beta(2) \end{vmatrix} - \begin{vmatrix} \psi_A(1)\beta(1) & \psi_A(1)\alpha(1) \\ \psi_A(2)\beta(2) & \psi_A(2)\alpha(2) \end{vmatrix}$$

$$\psi = 2\begin{vmatrix} \psi_A(1)\alpha(1) & \psi_A(1)\beta(1) \\ \psi_A(2)\alpha(2) & \psi_A(2)\beta(2) \end{vmatrix}$$

If the two electrons have the same spin, ($\alpha = \beta$), the determinant vanishes—that is, the wave function is zero, which is required by the exclusion principle if all the quantum numbers are the same.

Another way of stating the exclusion principle is that electrons having the same spin tend to avoid each other and do not occupy the same space. This property, termed spin correlation, is a guide to the position of electrons around an atom.[1,2] Consider, for example, the probable position of the two electrons in the excited state of helium $1s^1 2p^1$. Adequate one-electron space wave functions for the helium atom are

$$\psi_S = \frac{1}{\sqrt{2}}[\psi_{1s}(1)\psi_{2p}(2) + \psi_{1s}(2)\psi_{2p}(1)]$$

$$\psi_A = \frac{1}{\sqrt{2}}[\psi_{1s}(1)\psi_{2p}(2) - \psi_{1s}(2)\psi_{2p}(1)] \tag{4.4}$$

where S and A stand for symmetric and antisymmetric wave functions. The one-electron wave functions to be inserted in Equations (4.4) are

$$\psi_{1s} = f_{1s}(r_1)$$

$$\psi_{2p} = f_{2p}(r_2)\cos\theta$$

where r_1 and r_2 are the distances from the nucleus to electrons 1 and 2, respectively, θ (θ_1 and θ_2) is the angle between the line joining the electron and nucleus and a fixed line around which the wave function is symmetric, and $f_{1s}(r_1)$ and $f_{2p}(r_2)$ are the radial wave functions for the 1s and 2p orbitals, respectively.

Equations (4.4) can be rewritten

$$\psi_S = f_{1s}(r_1)f_{2p}(r_2)\cos\theta_2 + f_{1s}(r_2)f_{2p}(r_1)\cos\theta_1$$
$$\psi_A = f_{1s}(r_1)f_{2p}(r_2)\cos\theta_2 - f_{1s}(r_2)f_{2p}(r_1)\cos\theta_1$$

Assume the electrons are at the same distance from the nucleus; i.e., $r_1 = r_2 = r$. Then

$$\psi_S = f_{1s}(r)f_{2p}(r)(\cos\theta_2 + \cos\theta_1)$$
$$\psi_A = f_{1s}(r)f_{2p}(r)(\cos\theta_2 - \cos\theta_1)$$

(4.5)

When $\theta_1 = \theta_2 = 0$ or $\theta_1 = \theta_2 = \pi$, ψ_S^2 (the probability of finding the electrons) is a maximum, and ψ_A^2 is a maximum when $\theta_1 = 0$, $\theta_2 = \pi$, or vice versa. This means that for the singlet state ψ_S (spins antiparallel), the electrons are together, whereas for the triplet state (spins parallel), the electrons are on opposite sides of the nucleus.

Consider next beryllium, with the electronic configuration $(1s)^2(2s)^2$. In the excited state $(1s)^2(2s)^1(2p)^1$, it is clear from the preceding that the two valence electrons are expected to be at 180° to each other and that monomeric beryllium compounds BeX_2 should be linear, as, in fact, they are.

4.2   SPIN AND CHARGE CORRELATION[3]

It is important at this point to make a distinction between two terms: spin correlation, already mentioned, and charge correlation. Some confusion exists over the term "electron correlation," and it is not always clear what property of the electron, spin or charge, is being considered.

Spin correlation is the tendency for electrons with the same spin to keep apart while electrons with opposite spin tend to come together. Charge correlation is the tendency for particles of like charge to keep apart because of mutual repulsion. Therefore, spin and charge correlation work together for electrons with the same spin, but in opposition for electrons of opposed spins. To see how this is so, take eight uncharged particles, four of one spin and four of another spin. Spin correlation, which is the only effect operative here, places the particles in pairs with opposed spins at the corners of a tetrahedron [Fig. 4.1(a)]. However, if the particles all have the same charge, the additional effect of charge correlation will be to place the particles singly at the corners of a cube [Fig. 4.1(b)]. The net result is both spin and charge correlation within each tetrahedron of particles with parallel spins but only charge correlation between the two tetrahedrons of particles with opposed spins. This situation applies for the eight valence electrons of entities such as O^{2-}, F^-, and Ne. However, when such an atom is involved

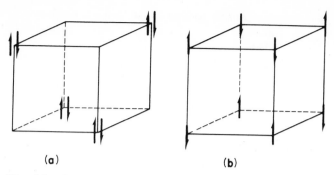

Fig. 4.1 Spin correlation (a) and spin and charge correlation (b) for eight particles.

in a compound, the bonding brings about some or complete correlation between the two tetrahedrons because of the localization of electron pairs in bonds. Consider the water molecule and assume, for the purposes of discussion, that the O^{2-} ion is approached by two protons. The formation of the two O—H bonds overcomes the charge correlation of the two electron pairs in the bonds, and the remaining electrons become more strongly correlated (Fig. 4.2). In the case of ammonia, where three corners of the tetrahedron are tied down, the electrons become completely correlated.

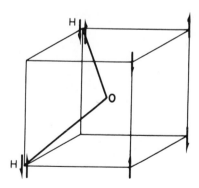

Fig. 4.2   Water showing electron correlation.

4.3   SPIN AND CHARGE CORRELATION AND STEREOCHEMISTRY

The concept of both charge and spin correlation can be used to predict bond angles in molecules such as H_2O and NH_3. Since the correlated electrons are at tetrahedral positions, the bond angles are expected to be

Table 4.1　　Stereochemical Arrangement for Two to Nine Electron Pairs

Number of electrons with the same spin *or* number of electron pairs	Shape	Example
2	Linear	$BeCl_2$ (gas)
3	Planar triangle	BF_3
4	Tetrahedron	CF_4
5	Trigonal bipyramid	PCl_5 (gas)
6	Octahedron	SF_6
7	Pentagonal bipyramid *or* irregular octahedron	IF_7
8	Square antiprism *or* dodecahedron	$Pb(acetate)_4$
9	Tripyramid (trigonal prism + one out of each face)	$\begin{cases} Nd(H_2O)_9^{3+} \\ ReH_9^{2-} \end{cases}$

close to the tetrahedral angle. Using spin and charge correlation, Table 4.1 gives the predicted shapes of molecules for different numbers of valence electrons.

This rationalization[2,4,5,6] of stereochemistry has considerable success when we view the gross stereochemistry of nontransition metal compounds. We shall now consider its application to a number of compounds.

PCl_5 Gas

The phosphorus atom has five valence electrons, which, on the basis of spin and charge correlation, will be found at the corners of a trigonal bipyramid. Each one of the electrons is then shared with an electron on a chlorine atom to give PCl_5. The molecular shape is therefore determined by the valence electron arrangement on the phosphorus atom. The 10 bonding electrons in the molecule are correlated irrespective of electron repulsions, as they are all involved in bonds.

ICl_4^- Ion

Iodine has seven valence electrons, but only four are used in bonding the chlorine atoms in ICl_4^-. Since the anion is diamagnetic, the four nonbonding electrons (a total of four because the species has one negative charge) must be two of one spin and two of the opposed spin. Hence, for ICl_4^- six electron pairs are completely correlated. They will be

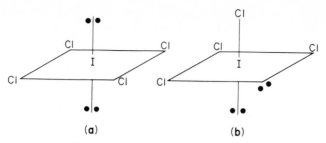

Fig. 4.3   Possible structure for ICl_4^-.

found at the corners of an octahedron, and the possible arrangements of bonding and nonbonding electron pairs are given in Fig. 4.3(a) and (b).

It is now necessary to consider qualitative estimates of the various electron repulsions possible for the two structures. The electrons involved in the bonds are shared between two atoms and the charge cloud is rather elongated. On the other hand, the lone pair is entirely attached to the central atom and is therefore closer to that atom than are the bonding pairs. Lone pairs hence occupy more space around the central atom than bonding pairs (Fig. 4.4).[7] The qualitative result is that repulsions between lone pairs will be greater than between bond pairs, and the order of decreasing repulsion energies is as follows: lone pair-lone pair repulsions > lone pair-bond pair repulsions > bond pair-bond pair repulsions. The various repulsions in ICl_4^- are

	Fig. 4.3(a)	Fig. 4.3(b)	Most stable
LP-LP	1 at 180°	1 at 90°	Fig. 4.3(a)
LP-BP	8 at 90°	6 at 90°	
		2 at 180°	Fig. 4.3(b)
BP-BP	4 at 90°	5 at 90°	
	2 at 180°	1 at 180°	Fig. 4.3(a)

The last column in the table is arrived at by assuming that the electron pair repulsions fall off rapidly with distance and that only short-range interactions are important, i.e., 90° interactions > 180° interactions. A qualitative estimate suggests that the arrangement in Fig. 4.3(a) is the most likely structure, which is correct.

H_2O

An investigation[8] of the various repulsions possible in a molecule can be used to calculate the equilibrium bond angle. Such an investigation shows clearly the importance of lone pairs. For example, for the water

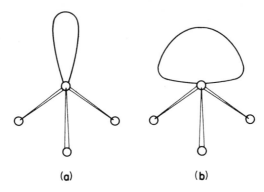

(a)   (b)

Fig. 4.4   Representation of (a) bond pair and (b) lone pair.

molecule, the various interactions are electron-electron repulsions in an orbital, electron-nucleus interactions, and various nonbonded interactions. In the third case, the interactions are nucleus-nucleus repulsions, electron-nucleus interactions between different bonds, and electron-electron interactions in different orbitals. The last term is the one of interest at present; it is made up of three parts:

E (bond-bond) repulsions

E (bond-lone pair) repulsions

E (lone pair-lone pair) repulsions

Table 4.2 shows how these terms vary with bond angle. As the bond angle is increased, the bond-bond repulsions drop, whereas the lone pair-lone pair repulsions increase. However, the lone pair-bond pair repulsions are at a minimum around 107°, where, in fact, the bond and lone pairs are at their maximum distance apart. The net result is a minimum energy in the vicinity of from 105 to 107° close to the actual bond angle. As the bond

Table 4.2[a]   Variation of H_2O Bond Angle with Various Bond-Lone Pair Interactions

$\widehat{HOH}$	E(bond-bond)	E(lone pair-lone pair)	E(bond-lone pair)	E(total)
90°	0.7664	2.8416	6.5226	10.131
100°	0.7683	3.0877	6.1928	10.049
105°	0.7588	3.1237	6.1410	10.024
109.5°	0.7526	3.1458	6.1350	10.033
120°	0.7344	3.1718	6.1646	10.071
180°	0.6660	3.1779	6.3941	10.238

angle is increased beyond 107°, both the lone pair-lone pair and lone pair-bond pair repulsions increase, preventing the formation of a linear molecule.

TeCl$_4$

We shall now consider an example where the most stable structure is not readily apparent. Tellurium in TeCl$_4$ has five electron pairs in its valence shell, which will be arranged at the corners of a trigonal bipyramid. The possible bond pair-lone pair arrangements are shown in

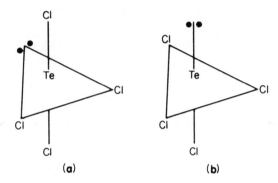

Fig. 4.5 Possible structures for TeCl$_4$.

Fig. 4.5(a) and (b). The various repulsions in these structures are listed as follows:

	Fig. 4.5(a)	Fig. 4.5(b)
LP-BP	2 at 90° 2 at 120°	3 at 90° 1 at 180°
BP-BP	4 at 90° 1 at 120° 1 at 180°	3 at 90° 3 at 120°

If we only consider the greatest repulsions, we can decide that Fig. 4.5(a) is the most likely arrangement. However, we cannot be completely sure of such qualitative predictions and especially of the importance of other repulsions that have been ignored. A more satisfactory method is to consider the space occupied by the lone pair.[9] Since lone pairs lie close to the central atom, the most stable configuration would probably be where the solid angle subtended by the lone pair is as large as possible. This is achieved for the structure represented in Fig. 4.5(a).

The molecule ClF_3 can be treated in the same way as $TeCl_4$. However, estimates of the total electrostatic repulsion energy between electron pairs indicate that the energy is insensitive to bond and lone pair arrangements. Results of calculations for ClF_3 given in Table 4.3 demonstrate that any arrangement is as good as any other.

Table 4.3[9]   Influence of Bond and Lone Pairs on Total Electro-
                static Energy of a Trigonal Bipyramidal Structure*

Polar	Equatorial	Polar/Equatorial distance	Total electrostatic energy (relative)
$2X$	$1X2L$	1	3.166
$1X1L$	$2X1L$	1	3.209
$2L$	$3X$	1	3.174
$2X$	$1X2L$	1.05	3.114
$1X1L$	$2X1L$	1.05	3.143
$2L$	$3X$	1.05	3.096

* X = Bond pair; L = Lone pair.

When using the described approach to stereochemistry, we do not need to consider every molecule separately. For example, $TeCl_4$ is an example of an AX_4L molecule, ClF_3 of an AX_3L_2 molecule, and ICl_2^- of an AX_2L_3 molecule, where L stands for lone pair. The predicted shapes for a number of molecular types are given in Table 4.4. The bond angles listed in the column head entitled Predicted Bond Angle are only a guide to the gross stereochemistry.

The use of the Pauli exclusion principle is less successful in explaining trends in bond angles for a related series of compounds. The decrease in the HXH angle (Table 4.5) from CH_4 to NH_3 to OH_2 appears related to an increase in the number of lone pairs that will push the bonding orbitals together. However, the big difference in bond angle between the hydrides of the first short-period elements and the hydrides of the succeeding elements is less readily explained.

Many proposals have been put forward to account for this change, such as electronegativity differences between the elements, or the ability of the heavier elements to expand their valence shell and use d orbitals. A simpler explanation is obvious from considering the covalent radii of the central atoms (Table 4.6) in the hydrides. Again, a marked change occurs in moving from the first short period to the heavier elements. The big decrease in bond angle could be simply a steric effect. The bonding electrons are further apart for the larger atoms, and consequently the bond angle can

Table 4.4   Stereochemistry Based on Pauli Exclusion Principle

Molecular species	Electron pairs	Lone pairs	Distribution of electron pairs		Molecular shape	Predicted bond angle	Examples
			Polar	Equatorial			
AX_2	2	0			Linear	180°	$BeCl_2$, gas
AX_3	3	0			Trigonal plane	120°	BX_3, $X = F$, Cl, Br, I
AX_2L	3	1			Bent	120°	SnX_2 gas
AX_4	4	0			Tetrahedron	109° 28'	CH_4, NH_4^+, SiF_4
AX_3L	4	1			Pyramidal	109° 28'	NH_3, H_3O^+, NF_3
AX_2L_2	4	2			Bent	109° 28'	H_2O, SCl_2, NH_2^-
AX_5	5	0	2X	3X	Trigonal bipyramid	90°, 120°	PCl_5 gas
AX_4L	5	1	1X, 1L	3X	Trigonal pyramid	90°, 120°	No examples
			2X	2X, 1L	Distorted tetrahedron	90°, 120°	$TeCl_4$, SbF_4^-, Me_2TeCl_2
AX_3L_2	5	2	2L	3X	Trigonal plane	120°	No examples
			1L, 1X	1L, 2X	Trigonal pyramid	90°, 120°	No examples
			2X	2L, 1X	T-shaped	90°	ClF_3, BrF_3
AX_2L_3	5	3	2L	1L, 2X	Bent	120°	No examples
			1L, 1X	2L, 1X	Bent	90°	No examples
			2X	3L	Linear	180°	ICl_2^-, I_3^-, XeF_2
AX_6	6	0			Octahedron	90°	SF_6, PCl_6^-, AlF_6^{3-}
AX_5L	6	1			Square pyramid	90°	SbF_5^{2-}, IF_5, BrF_5
AX_4L_2	6	2	2L trans		Square	90°	ICl_4^-, XeF_4
			2L cis		Distorted tetrahedron	90°	No examples
AX_7	7	0	2X	5X	Pentagonal bipyramid	90°, 72°	IF_7
AX_6L	7	1	1X, 1L	5X	Pentagonal pyramid	90°, 72°	$SbBr_6^{3-}$
			2X	4X, 1L	Distorted octahedron	90°	No examples XeF_6?
AX_8	8	0			Square antiprism		TaF_8^{3-}, $Th(C_5H_7O_2)_4$
					Dodecahedron		$Mo(CN)_8^{4-}$
AX_9	9	0			Tripyramid		$Nd(H_2O)_9(BrO_3)_3$

Table 4.5   Bond Angles

CH_4	NH_3	OH_2
109.5°	107.3°	104.5°
	PH_3	SH_2
	93.3°	92.2°
	AsH_3	SeH_2
	91.8°	91.0°
	SbH_3	TeH_2
	91.3°	89.5°

Table 4.6   Covalent Radii (pm)*

N	70	O	66
P	110	S	104
As	118	Se	114
Sb	136	Te	132

* pm = picometer, 10^{-12} m., $1\text{Å} = 10^2$ pm.

decrease before repulsions become as great as for hydrides of nitrogen and oxygen.

The preceding procedure of calculating electron pairs and their position in space ignores any consideration of the bonding orbitals that may be used by the central atom. For example, the five electron pairs in the valence shell of phosphorus in PCl_5 would need to be accommodated in molecular orbitals having contributions from the $3s$, $3p$, and $3d_{z^2}$ orbitals of phosphorus. However, it is questionable whether the frequent assumption that d orbitals are involved in the bonding, in such compounds, is justified. Therefore, it is important to realize that the approach outlined is a method to describe and predict the broad features of shape; for this reason, it is a useful tool. The method by which the central element achieves this through its bonding orbitals is a more difficult problem, as Chapters 3 and 5 outline.

4.4 THE PAULI PRINCIPLE AND HIGH COORDINATION NUMBERS

Special problems are associated with seven or more electron pairs in the valence shell, owing, in part, to the similar energies of different stereochemical arrangements.

An interesting situation arises for the seven-electron pair system AX_6L found in SeX_6^{2-}, TeX_6^{2-} ($X = Cl$, Br), $SbBr_6^{3-}$ and XeF_6. According to the data in Table 4.3, the seven electron pairs are at the points of either a pentagonal bipyramid or a distorted octahedron. The antimony complex anion $SbBr_6^{3-}$ has a distorted octahedral structure, but the selenium and tellurium complexes have a cubic space group and are isomorphous with the transition metal halogeno-complex anions MX_6^{2-} ($M = Re$, Os, Ir, Pt). This means that the lone pair in the case of selenium and tellurium is not stereochemically active. It does not appear to be involved in any multiple bonding, because the Te—Cl bond is 251 pm in $TeCl_6^{2-}$, whereas it is 233 pm in gaseous $TeCl_4$ (where pm is picometer, 10^{-12} m; 1 angstrom unit $\text{Å} = 10^2$ pm). This anomaly may or may not extend to XeF_6, whose structure is still in doubt, although most evidence suggests that XeF_6 has a distorted octahedral structure. Seven stereochemically active electron pairs around xenon seem likely, in view of the ability of xenon to expand its valence shell, to give the complex fluoroanions XeF_7^- and XeF_8^{2-}.[6]

If, for steric reasons, it is difficult to place seven electron pairs around selenium and tellurium, then fluoroanions would provide the situation of least steric interaction. Interestingly, both elements form the fluoro-complexes MF_5^-, suggesting a great reluctance of selenium and tellurium to accommodate seven electron pairs, all stereochemically active. However, the lone pair must be somewhere, and the uniformly long Te—Cl bond in $TeCl_6^{2-}$ suggests that it is in a spherically symmetrical orbital, repelling all six ligands equally.

There is no doubt as to the most stable electron arrangement for up to six electron pairs, but for seven and eight electron pairs, the situation is not so clear.[6,10] The problem can be approached by considering the force between the electron pairs. For seven particles (with the same spin), the possible arrangements are shown in Fig. 4.6. Actual examples are known

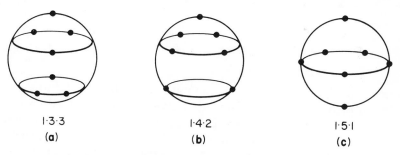

1·3·3   1·4·2   1·5·1

(a)   (b)   (c)

Fig. 4.6 Arrangement of seven particles: (a) 1.3.3, octahedron plus a particle opposite one face; (b) 1.4.2, trigonal prism plus a particle opposite one face; (c) 1.5.1, pentagonal bipyramid.

for each structure, indicating that the energies of the three arrangements are probably similar.

For eight electron pairs, the two principal arrangements are given in Fig. 4.7. The cube is less stable than the square antiprism or dodecahedron, but there is probably little to choose between the latter two.

Of all the possible arrangements for nine electron pairs, the tripyramid (Fig. 4.8) is the most stable and is so far the only stereochemistry found for nine coordination.

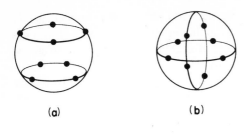

(a)   **(b)**

Fig. 4.7 Arrangement of eight particles: square antiprism; (b) dodecahedron.

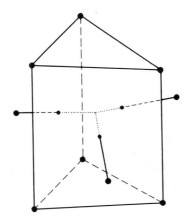

Fig. 4.8 Tripyramid.

4.5 AX$_5$ AND AX$_7$ SPECIES[6,11]

The molecular or ionic types AX_5 and AX_7, which have the trigonal bipyramidal or pentagonal bipyramidal structures, respectively, do not have all the ligand positions equivalent, i.e., the polar positions are not symmetry related to the equatorial positions. For example, for the trigonal bipyramid the polar-equatorial bond angle is 90°, whereas the equatorial-equatorial angle is 120°. In addition, each polar position experiences three repulsions at 90° and one at 180°, whereas each equatorial position experiences two at 120° and two at 90°. Since the repulsive forces fall off rapidly with distance, it would be expected that the polar ligands will experience greater repulsion than will the equatorial ligands. To offset the imbalance of forces, the polar $A—X$ bond lengths are somewhat longer than the equatorial bond lengths. This is observed for a number of AX_5, AX_4L, AX_3L_2 molecules and the one $AX_7(IF_7)$ molecule whose structure is known (Table 4.7). We may expect that the lone pairs in the equatorial positions would further increase the polar $A—X$ bond lengths owing to the greater repulsive effect of the lone pairs; however, there is no evidence yet to indicate a greater bond length in AX_4L and AX_3L_2 species compared with AX_5 species.

Table 4.7 Stereochemistry of AX_5, AX_4L, AX_3L_2, AX_5L, AX_6L, and AX_7 Species

Molecular species	Compound	Central atom radius (pm)			Bond angles		Shape
		Equatorial (E)	Polar (P)	Difference (pm)	PAE	EAE	
AX_5	PCl_5	105	120	15			
	PF_5	89	94	5	90°	120°	
	PPh_5	108	122	14			
	CH_3PF_4	90	97	7	FPC 91.8°	FPC 122.2°	
	$(CH_3)_2PF_3$	91	100	9	FPF 88.9°	FPC 118°	
	PCl_2F_3	95	106	11			
	$SbCl_5$	132	144	12			
	$(C_2H_2Cl)_3SbCl_2$	146	138	8			
AX_4L	SF_4	91	101	10			
	SOF_3	90	96	6			
	Ph_2SeBr_2	114	136	22	90°	110°	
	Ph_2SeCl_2	114	131	17	90°		
	$(tol)_2SeCl_2$*	117	140	23	91.5°	108°	
	$(tol)_2SeBr_2$*	117	140	23	91°	106.5°	
	Me_2TeCl_2	133	152	19	86.2°	98.2°	
	Ph_2TeBr_2	137	154	17			
	$IO_2F_2^-$	127	136	9	90°	100°	
AX_3L_2	ClF_3	94	104	10	87.5°		
	$PhICl_2$	123	146	23	86.2°		
	BrF_3	108	118	10	86°		
AX_5L	$SbCl_5^{2-}$	262†	236†	26			
	SbF_5^{2-}	179†	208†	29			
	BrF_5	168†	202†	34			
AX_7	IF_7	183†	194†	11			
AX_6L	$SbBr_6^{3-}$						

* tol $= CH_3-C_6H_4-$.   † $A-X$ bond lengths.

71

The six-electron pair species AX_5L, typified by $SbCl_5^{2-}$, SbF_5^{2-}, and BrF_5, have a square-based pyramidal structure with the lone pair in the sixth octahedral position. In this situation, some difference in bond lengths may be expected, but it is not easy to predict which bonds will be lengthened. The ion $SbCl_5^{2-}$ has its equatorial bonds lengthened, but for SbF_5^{2-} and BrF_5 the polar bond is longest.

Compounds such as CH_3PF_4, $(CH_3)_2PF_3$, $Ph_2SeX_2(X = Cl, Br)$ have the less electronegative groups in the equatorial positions. Presumably, steric factors and size of the bonding orbitals of the ligands determine their position in space.

4.6 MULTIPLE BONDS[2,6]

The Pauli principle can be used to rationalize the stereochemistry of non-transition element compounds that also have some multiple bonding. If the σ and π electrons between two atoms are considered to occupy the same space, then the gross shape of a molecule can be determined by considering the arrangement of σ and lone-pair electrons, as before. Finer detail is then derived by considering the relative repulsions imposed by triple, double, and single bonds. Since four electrons in a double bond will occupy more space than two electrons in a single bond, a qualitative order of some repulsions is as follows: triple bond-triple bond repulsions > triple bond-double bond > double bond-double bond > double bond-single bond > single bond-single bond repulsions. Table 4.8 lists the various shapes for different numbers of electron pairs in molecules with multiple bonds. The effect of double bonds on the finer details of stereochemistry is apparent from Table 4.9.

4.7 AN ELECTROSTATIC MODEL[12,13]

If we assume that each bonded or lone pair of electrons has a characteristic charge concentrated along the bond direction or the direction the lone pair points, and if these charges are taken as equidistant from the central atom, by a process of minimizing the electrostatic interaction we can determine the shape of a molecule. For a molecule $AXYZ$,

$$Z-A\begin{matrix} \nearrow X \\ \searrow Y \end{matrix}$$

Table 4.8 Shapes of Molecules with Multiple Bonds

Total number of σ bonds and lone pairs	Arrangement	σ Bonds	Lone pairs	π Bonds	Molecular shape	Examples
2	Linear	2	0	2	Linear	$O=C=O$, $H-C\equiv N$, $O=N=O^+$
3	Triangular plane	3	0	1, 3	Triangular plane	$Cl_2C=O$, HCO_2^-, SO_3, NO_3^-
		2	1	1, 2	Bent	SO_2, $ClN=O_2$, $O=N-O^-$
4	Tetrahedral	4	0	1, 2	Tetrahedron	$O=PCl_3$, SO_2Cl_2, SO_4^{2-}
		3	1	1	Pyramidal	$SOCl_2$, ClO_3^-
		2	2	1	Bent	ClO_2^-
5	Trigonal bipyramid	5	0	1	Trigonal bipyramid	No examples
		4	1	1	Irregular tetrahedron	$IO_2F_2^-$
6	Octahedron	6	0	1	Octahedron	H_5IO_6

73

Table 4.9 Influence of Double Bonds on Stereochemistry

Molecular type	Molecules	Angle *sb-sb* (deg)*	Angle *sb-db* (deg)*
AX_2Y	$F_2C{=}O$	112.5	123.2
	$Cl_2C{=}O$	111.3	124.3
	$H_2C{=}O$	118	121
	$(NH_2)_2C{=}O$	118	121
	$(CH_3)_2C{=}CH_2$	109	125
	Expected	< 120	> 120
AX_2YL	$F_2S{=}O$	92.8	106.8
	$Br_2S{=}O$	96	108
	$(CH_3)_2S{=}O$	100	107
	$(C_6H_5)_2S{=}O$	97.3	106.2
	$Cl_2Se{=}O$	106	114
	Expected	< 109.5	> 109.5
AX_3Y	$F_3P{=}O$	102.5	
	$Cl_3P{=}O$	103.6	
	$Br_3P{=}O$	108	
	$F_3P{=}S$	100.3	
	$Cl_3P{=}S$	100.5	
	$Br_3P{=}S$	106	
	Expected	< 109.5	
			db-db (deg)
AX_2Y_2	F_2SO_2	96.1	124
	Cl_2SO_2	112.2	119.8
	$(NH_2)_2SO_2$	112.1	119.4
	$(CH_3)_2SO_2$	115	125

* *sb* = Single bond.
  db = Double bond.

the electrostatic repulsion energy is given by

$$E_T = \frac{XY}{2d \sin \dfrac{\omega}{2}} + \frac{YZ}{2d \sin \dfrac{\psi}{2}} + \frac{ZX}{2d \sin \dfrac{\varphi}{2}}$$

where $\omega = X\hat{A}Y$, $\psi = Y\hat{A}Z$, and $\varphi = Z\hat{A}X$, d is the distance from the central atom to the characteristic charge, and X, Y, and Z are the characteristic charges. From a knowledge of the charges and by minimizing the energy given by the preceding equation, we can find the bond angles. We can do so empirically by finding a quantity called the electrostatic repulsion number, ERN, which measures the repulsion between electrons bonded to various atoms compared with a lone pair taken as unity. The ERN values

Table 4.10 Electrostatic Repulsion Numbers

Element	ERN	Comment
O	0.94 ⎱	
S	0.94 ⎰	Bonded to all elements.
Cl	0.72 ⎫	
F	0.63 ⎬	Bonded to second period element.
H	0.81 ⎭	
All halogens	0.52 ⎫	
H	0.38 ⎬	Bonded to third or higher period element.
CH_3	0.93	

are obtained for bonded atoms from molecules with known bond angles. Typical values are given in Table 4.10; such values can now be used in predicting bond angles in other compounds, with reasonable success (see Table 4.11).

Table 4.11

Molecule		Predicted bond angle	Experimental bond angle
IBr_2^-		180°	180°
$IClBr^-$		180°	180°
I_3^-		180°	180°
$IO_2F_2^-$	$\widehat{FIF}$	180°	180°
	$\widehat{OIO}$	118°	100° ± 10°
ClF_3	$\widehat{F_pClF_{eq}}$	79°	87° 12.6′
BrF_3	$\widehat{F_pClF_{eq}}$	79°	86° 12.6′
SeF_4		137.5°	120°
$TeCl_4$		83°	93° ± 3°
ICl_4^-		90°	90°
BrF_4^-		90°	90° or 109.5°
BrF_5	$\widehat{F_pBrF_{eq}}$	82.4°	84° ± 3.5°
CH_3OH	$\widehat{COH}$	106°	107° 16′
CH_3NH_2	$\widehat{HNH}$	105° 30′	105° 50′
H_2O_2	$\widehat{HOO}$	100°	97°
S_2Cl_2	$\widehat{ClSS}$	105°	107°
$S_2O_4^{2-}$	$\widehat{SSO}$	101°	99°
	$\widehat{OSO}$	109°	108°
C_2Cl_4	$\widehat{ClCCl}$	110° 30′	113° 18′

4.8   CONFIGURATION OF MAXIMUM PROBABILITY

Using a technique devised by Zimmerman and Rysselberghe,[14] Linnett and his coworkers[15–18] have determined the configuration of maximum probability for the elements of the first short period. If we consider all the electrons, the angle between the unpaired electrons of oxygen is 103°, which is close to the observed angle in OX_2 compounds

The method is to write down the wave function for the system under consideration, taking into account all the electrons. The probability function ψ^2 is dependent on the spatial coordinates of the electrons; hence, a parameter such as bond angle is selected to make this function a maximum. A number of configurations of peak probability are found for each atom, one of which corresponds with the configuration of maximum probability. The results for the first short period are summarized in Table 4.12.

The method is less successful for the second short period elements. For example, the HPH bond angle in PH_3 is predicted to be 109° 40′, which is wrong. Clearly, additional features, such as the use of d orbitals, may be important here. However, the method has also been used to predict the correct shapes for $Ni(CN)_4^{2-}$, SF_6, $Fe(CN)_6^{4-}$, PF_5, and $Mo(CN)_8^{4-}$.

4.9   SUMMARY

In this chapter, a number of approaches to stereochemistry have been outlined, but in all cases they depend on the fundamental requirements of the Pauli exclusion principle. The success of this principle in describing the shape of molecules lends strong support to the generally held idea that the gross stereochemistry displayed by the elements is largely determined by the outer valence electrons. The theories differ in their consideration of the finer detail of other effects, such as electron repulsion, ligand interactions, and the orbitals available for bonding.

BIBLIOGRAPHY

1. M. C. DAY and J. SELBIN, *Theoretical Inorganic Chemistry*, Reinhold Publishing Corp., New York, 1962.

2. R. J. GILLESPIE and R. S. NYHOLM, *Quart. Rev. London*, **11**, 339 (1957).

3. J. W. LINNETT, *The Electronic Structure of Molecules*, Methuen & Co., Ltd., London, 1964.

Table 4.12[15] Configurations of Maximum Probability*

Atom	B(^{2}P)	C(^{3}P)	N(^{4}S)	O(^{3}P)	F(^{2}P)	Ne(^{1}S)	Be(^{3}P)	B(^{4}P)	C(^{5}S)
Configuration	P—B—S	C(P, S, S)	N(P, S, S, S)	O(P, P, S, S)	F(P, P, P, S)	N(P, P, P, P)	S—Be—S	B(S, S, S)	C(S, S, S, S)
Symmetry	$C_{\infty v}$	C_{2v}	C_{3v}	C_{2v}	C_{3v}	T_d	$D_{\infty h}$	D_{3h}	T_d
$\widehat{SES}$		118°	108°	103°	101°		180°	120°	109.5°
$\widehat{PES}$	180°	121°	111°	104.5°	116°				
$\widehat{PEP}$				133.5°		109.5°			

* P corresponds with paired electrons; S corresponds with unpaired electrons.

4. N. V. SIDGWICK and H. POWELL, *Proc. Roy. Soc. London, Ser. A*, **176**, 153 (1940).

5. J. E. LENNARD-JONES and J. A. POPLE, *Phil. Mag.*, **43**, 581, 1953.

6. R. J. GILLESPIE, *Angew. Chem.* (Int. Ed.), **6**, 819 (1967), and references therein.

7. R. J. GILLESPIE, *J. Am. Chem. Soc.*, **82**, 5978 (1960).

8. J. A. POPLE, *Proc. Roy. Soc. London, Ser.* **202A**, 323 (1950).

9. A. SEARCY, *J. Chem. Phys.*, **31**, 1 (1959).

10. R. J. GILLESPIE, *Can. J. Chem.*, **38**, 818 (1960).

11. R. J. GILLESPIE, *Can. J. Chem.*, **39**, 318 (1961).

12. A. SEARCY, *J. Chem. Phys.*, **28**, 1237 (1958).

13. A. PARSONS and A. SEARCY, *J. Chem. Phys.*, **30**, 1635 (1959).

14. H. K. ZIMMERMAN and P. VAN RYSSELBERGHE, *J. Chem. Phys.*, **17**, 598 (1949).

15. J. W. LINNETT and A. J. POE, *Trans. Faraday Soc.*, **47**, 1033 (1951).

16. C. E. MELLISH and J. W. LINNETT, *Trans. Faraday Soc.*, **50**, 657 (1954).

17. J. W. LINNETT and C. E. MELLISH, *Trans. Faraday Soc.*, **50**, 665 (1954).

18. P. G. DICKENS and J. W. LINNETT, *Trans. Faraday Soc.*, **53**, 1037 (1957).

5

Molecular Orbital Theory

5.1 DIATOMIC MOLECULE

The molecular orbital (MO) theory is conceptually more simple than the valence bond theory, as it is basically an extension of the quantum theory of atoms. Linear combinations of atomic orbitals (LCAO) are taken in order to form molecular orbitals for the molecule into which electrons are fed in the usual way.

For a homonuclear diatomic molecule AB, the molecular orbital wave function is given by the equation

$$\psi_{MO}^{\pm} = c_1 \psi_A \pm c_2 \psi_B \tag{5.1}$$

where ψ_A and ψ_B are the wave functions for the bonding atomic orbitals of atoms A and B, respectively. The energy of the molecular orbitals is determined by minimizing the energy in Equation (5.2)

$$\hat{H}\psi_{MO}^{\pm} = E\psi_{MO}^{\pm} \tag{5.2}$$

with respect to the two parameters c_1 and c_2, the coefficients in the molecular orbital wave functions. The result[1] of this procedure is

$$E^{\pm} = \frac{H_{AA} \pm H_{AB}}{1 \pm S_{AB}} \tag{5.3}$$

79

where the Coulomb integral is $H_{AA} = \int \psi_A \hat{H} \psi_A \, dT$ or H_{BB}, the resonance integral is $H_{AB} = H_{BA} = \int \psi_A \hat{H} \psi_B \, dT$, and the overlap integral is $S_{AB} = \int \psi_A \psi_B \, dT$.

The orbital energy is clearly dependent on the overlap integral, which is a function of the internuclear distance. The orbitals corresponding to the

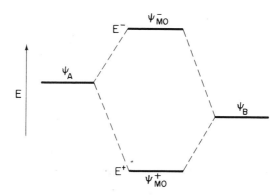

Fig. 5.1   Molecular orbital diagram for a diatomic molecule *AB*.

energies E^+ and E^- are ψ_{MO}^+ and ψ_{MO}^- and are termed bonding and antibonding, respectively. The result is represented diagrammatically in an orbital energy diagram (Fig. 5.1).

Table 5.1   Molecular Orbitals

Orbital	Rotation about internuclear axis	Number of nodes	Component of angular momentum about the internuclear axis
σ	Symmetrical	0	0
π	Antisymmetrical	1	1
δ	Antisymmetrical	2	2

The type of molecular orbitals formed are σ, π, and δ, the classification being according to the component of angular momentum about the internuclear axis (Table 5.1). The different molecular orbitals are depicted in Fig. 5.2.

It is now possible to write the molecular orbitals for diatomic homonuclear molecules, taking the z axis as the bond direction, (Table 5.2). The usual order of energies of the orbitals is:

$$\sigma^b 1s < \sigma^* 1s < \sigma^b 2s < \sigma^* 2s < \sigma_z^b 2p < (\pi_x^b 2p = \pi_y^b 2p)$$
$$< (\pi_x^* 2p = \pi_y^* 2p) < \sigma_z^* 2p$$

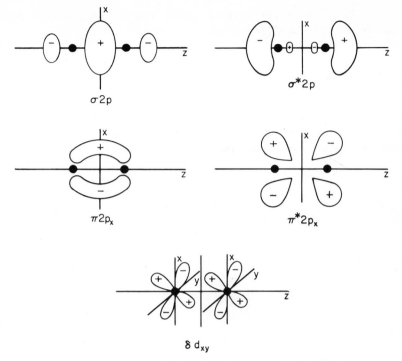

Fig. 5.2   Molecular orbitals, $\sigma 2p$, $\sigma^* 2p$, $\pi 2p_x$, $\pi^* 2p_x$, and δd_{xy}.

Table 5.2   Molecular Orbitals

Atomic orbitals		Molecular orbitals	
A	B		
$1s$	$1s$	$\sigma^b 1s$	$\sigma^* 1s$
$2s$	$2s$	$\sigma^b 2s$	$\sigma^* 2s$
$2p_z$	$2p_z$	$\sigma_z^b 2p$	$\sigma_z^* 2p$
$2p_x$	$2p_x$	$\pi_x^b 2p$	$\pi_x^* 2p$
$2p_y$	$2p_y$	$\pi_y^b 2p$	$\pi_y^* 2p$

although the order will vary with the internuclear distance. All the electrons are now fed into the orbitals in the usual way. For example,

$$N_2: \quad (\sigma^b 1s)^2 (\sigma^* 1s)^2 (\sigma^b 2s)^2 (\sigma^* 2s)^2 (\sigma_z^b 2p)^2 (\pi_{xy}^b 2p)^4$$

an effective triple bond.

The distribution of electrons in the molecular orbitals for oxygen is

$$(\sigma^b 1s)^2 (\sigma^* 1s)^2 (\sigma^b 2s)^2 (\sigma^* 2s)^2 (\sigma_z^b 2p)^2 (\pi_{xy}^b 2p)^4 (\pi_{xy}^* 2p)^2$$

which predicts that oxygen is paramagnetic with two unpaired electrons in the doubly degenerate π_{xy}^*2p orbitals.

5.2 POLYATOMIC MOLECULES

We shall now illustrate the MO procedure for the pyramidal molecule ammonia. The method of forming the molecular orbitals is simplified by the use of symmetry and group theory.[2,3]

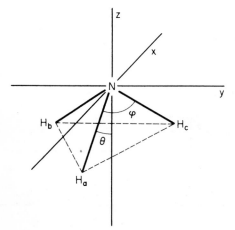

Fig. 5.3   Coordinate system for ammonia.

Consider the molecule oriented as in Fig. 5.3, where each NH bond makes angles θ with the z axis and φ with the other NH bonds. The NH$_a$ bond is taken to lie in the xz plane. The symmetry orbitals for the atomic orbitals of nitrogen in the C_{3v} point group are

$$1s, 2s, 2p_z \quad a_1$$

$$(2p_x2p_y) \quad e \quad \text{(see Table 2.11)}$$

The three NH bonds transform as the reducible representation

$$E \quad 2C_3 \quad 3\sigma_v$$

$$\Gamma_{N-H} \quad 3 \quad 0 \quad 1$$

which reduces to $A_1 + E$. Since we require only one nitrogen a_1 orbital, we shall omit the nitrogen $2s$ orbital from our initial discussion and assume the $1s$ orbital is not involved in the bonding.

To form molecular orbitals with the nitrogen $a_1(2p_z)$ and $e(2p_x, 2p_y)$ orbitals and the hydrogen $1s$ orbitals, we must find the *group orbitals* of the three hydrogen $1s$ orbitals. That is, find three combinations of the hydrogen $1s$ orbitals, one of which has the same transformation property as the a_1 representation and two of which have the same transformation properties as the e representation of the C_{3v} point group.

The hydrogen group orbitals in ammonia are generated by operation on one or more of the hydrogen $1s$ orbitals by the operations of the C_{3v} point group. The transformed orbital is then multiplied by the character of the irreducible representation being considered (A_1 or E) and the results summed. This will give unnormalized group orbitals. The transformation table for hydrogen $1s_a$, $1s_b$, and $1s_c$ orbitals is as follows:

	E	C_3	C_3^2	σ_{va}	σ_{vb}	σ_{vc}
$1s_a$	$1s_a$	$1s_b$	$1s_c$	$1s_a$	$1s_c$	$1s_b$
$1s_b$	$1s_b$	$1s_c$	$1s_a$	$1s_c$	$1s_b$	$1s_a$
$1s_c$	$1s_c$	$1s_a$	$1s_b$	$1s_b$	$1s_a$	$1s_c$

Multiply the terms of the first row of the table by the characters of the A_1 representation, and sum to give

$$\psi_{A_1} = 2(1s_a) + 2(1s_b) + 2(1s_c)$$

The normalized orbital is

$$\psi_{A_1} = \sqrt{\tfrac{1}{3}}(1s_a + 1s_b + 1s_c) \tag{5.4}$$

For the doubly degenerate orbitals (E_1), multiply the first row with the characters of the E representation,

$$\psi'_E = 2(1s_a) - 1s_b - 1s_c$$

which, on normalization, gives

$$\psi'_E = \frac{1}{\sqrt{6}}[2(1s_a) - 1s_b - 1s_c] \tag{5.5}$$

The second component of E is obtained by using the second row of the transformation table to give

$$\psi''_E = 2(1s_b) - 1s_c - 1s_a$$

Normalization gives

$$\psi''_E = \frac{1}{\sqrt{6}}[2(1s_b) - 1s_c - 1s_a] \tag{5.6}$$

To ensure that the ψ_E wave functions are orthogonal, keep ψ'_E as it is and find a new function $\psi'''_E = \psi''_E + C\psi'_E$, so that ψ'_E and ψ'''_E are orthogonal. The constant C is $-\tfrac{1}{2}$ for this to be so; hence,

$$\psi'''_E = \sqrt{\tfrac{1}{2}}(1s_b - 1s_c) \tag{5.7}$$

We are now in a position to write down the MO's for the NH_3 molecule.

$$\psi(\sigma_z^b) = C_1 2p_z + C_2 \frac{1}{\sqrt{3}}(1s_a + 1s_b + 1s_c) = 3a_1$$

$$\psi(\sigma_z^*) = C_3 2p_z - C_4 \frac{1}{\sqrt{3}}(1s_a + 1s_b + 1s_c) = 4a_1$$

$$\left. \begin{array}{l} \psi(\sigma_y^b) = C_5 2p_y + C_6 \frac{1}{\sqrt{2}}(1s_b - 1s_c) \\[4mm] \psi(\sigma_x^b) = C_7 2p_x + C_8 \frac{1}{\sqrt{6}}[2(1s_a) - 1s_b - 1s_c] \end{array} \right\} = 1e \qquad (5.8)$$

$$\left. \begin{array}{l} \psi(\sigma_y^*) = C_9 2p_y - C_{10} \frac{1}{\sqrt{2}}(1s_b - 1s_c) \\[4mm] \psi(\sigma_x^*) = C_{11} 2p_x - C_{12} \frac{1}{\sqrt{6}}[2(1s_a) - 1s_b - 1s_c] \end{array} \right\} = 2e$$

The molecular orbitals are labeled according to their symmetry based on group theory. The first orbital in Equation (5.8) (labeled a_1) is called $3a_1$ because it is the third most stable a_1 orbital of the complete molecule. The nitrogen $1s$ and $2s$ orbitals not considered previously are $1a_1$ and $2a_1$, respectively, in the molecule.

The wave functions (5.8) can now be used to determine the energy of the molecule. To do this, we have to evaluate the overlap integrals, called group overlap integrals.[2] For the $3a_1$ orbital, the group overlap integral is

$$S(3a_1) = \int 2p_z \frac{1}{\sqrt{3}}(1s_a + 1s_b + 1s_c)\, dT \qquad (5.9)$$

It cannot be evaluated as it stands, because its various terms, such as $\int 2p_z(1s_a)\, dT$, are not known because the $1s_a$ orbital does not lie along the z axis (Fig. 5.3). However, overlap integrals are tabulated for the overlap of $2p$ and $1s$ orbitals when the orbitals lie along the same axis. It is possible to express Equation (5.9) in terms of these known two-center overlap integrals $S(1s, 2p) = \int 1s2p\, dT$ as follows:

$$S(3a_1) = \frac{1}{\sqrt{3}} \int (2p_z 1s_a + 2p_z 1s_b + 2p_z 1s_c)\, dT$$

$$= \frac{1}{\sqrt{3}}[\cos \theta S(1s, 2p\sigma) + \cos \theta S(1s, 2p\sigma) + \cos \theta S(1s, 2p\sigma)] \qquad (5.10)$$

$$= \sqrt{3} \cos \theta S(1s, 2p\sigma)$$

and

$$S(1e)_y = \int 2p_y \frac{1}{\sqrt{2}}(1s_b - 1s_c)\, dT$$

$$= \frac{1}{\sqrt{2}} [\cos 30 \sin \theta S(1s, 2p\sigma) + \cos 30 \sin \theta S(1s, 2p\sigma)] \qquad (5.11)$$

$$= \sqrt{\frac{3}{2}} \sin \theta S(1s, 2p\sigma)$$

and

$$S(1e)_x = \sqrt{\frac{3}{2}} \sin \theta S(1s, 2p\sigma) \qquad (5.12)$$

In principle, it should be possible to find the value of θ, and hence φ, which minimizes the energy. However, as we shall see later, the energy is not very sensitive to changes in the value of the bond angle. If $\theta = 90°$, which corresponds to a planar molecule, it can be seen from Equation (5.10) that $S(3a_1) = 0$, whereas $S(1e)_x = S(1e)_y = \sqrt{3/2}S(1S, 2p)$, and by geometry $\varphi = 120°$. On the other hand, if $\varphi = 90°$, we can show that $\sin \theta = \sqrt{2/3}$ and $\cos \theta = \sqrt{3}/3$. Hence, $S(3a_1) = S(1e)_x = S(1e)_y = S(1s, 2p)$. This corresponds to the situation when the protons approach along the x, y, and z axes.

An improvement of the molecular orbitals can be obtained by including the nitrogen $2s(a_1)$ orbital [and even the nitrogen $1s(a_1)$ orbital] in the bonding scheme. These orbitals can be combined with the nitrogen $2p_z(a_1)$ orbital in the molecular wave functions, inasmuch as they have the same symmetry properties. In the case of the water molecule, which can be treated in the same way, it has been shown that inclusion of the $1s$ orbital of oxygen still gives a minimum energy corresponding to a bond angle greater than 120°. The calculated molecular energy is 2062.5 eV for a bond angle of 105°, whereas the experimental value of the energy is -2080.6 eV. In fact, it is found that changes in the bond angle have little effect upon the total energy of the molecule, as the following figures indicate.[4]

Bond angle (deg)	100	105	110	120	180
Total energy (eV)	-2062.3	-2062.5	-2062.8	-2063.0	-2062.0

On the other hand, the calculated dipole moment of $1.52D$ at 105° agrees with the experimental value.

Calculations using the self-consistent field method give much better results even though changes in energy are still insensitive to changes in bond angles. Results from a few calculations are given as follows:[5]

Molecule	H_2O	F_2O	Li_2O	O_3
Calculated	110°	102°	180°	118.5°
Experimental	104.5°	103°	180°	117°

Clearly, although it is relatively easy to determine that the water molecule is bent, it becomes more difficult to predict that the bond angle is 104.5°. Whereas this must be one of the ultimate aims of quantum chemistry, until it is achieved readily for a number of compounds we must be content with semiquantitative descriptions. Fortunately, such descriptions are informative when we discuss the shapes of molecules, particularly in comparing a related series of compounds.

5.3 STRUCTURE CORRELATION

It is instructive to consider the factors that determine why a molecule takes up one shape rather than another; for example, why is NH_3 pyramidal rather than planar, and why is the reverse true for the nitrate ion? One approach is to correlate between the molecular orbital wave functions for two limiting shapes.[6] The term "correlation" is used in the same way as when applied to correlation diagrams for homonuclear and heteronuclear di-atomic molecules.[7] From what has already been said, it is possible to write down the distribution of electrons in the molecular orbitals of the pyramidal structure of NH_3 as follows:

$$(1a_1)^2(2a_1)_b^2(1e)_b^4(3a_1)_{nb}^2 \qquad (5.13)$$

The electron distribution in the molecular orbitals of a planar MH_3 structure (D_{3h}) is

$$(1a_1')^2(2a_1')_b^2(1e')_b^4(1a_2'')_{nb}^2 \qquad (5.14)$$

The $2a_1$ and $3a_1$ orbitals of (5.13) are made up of the $2s$ and $2p_z$ orbitals of nitrogen. The $2a_1'$ orbital in (5.14) consists of only the $2s$ orbital of nitrogen, because the $2p_z$ orbital is nonbonding and becomes the $1a_2''$ orbital in the planar structure.[8] The s-p_z mixing in the $2a_1$ bonding orbital of the pyramidal structure means that it will be a stronger bonding orbital than the $2a_1'$ bonding orbital of the planar molecule. However, the $1e'$ orbital (5.14) is more strongly bonding than the $1e$ orbital (5.13), owing to the greater overlap in the former case. The overlap integral for the pyramidal molecule is $\sqrt{3/2}\sin\theta S(1s, 2p)$, whereas for a planar molecule it is $\sqrt{3/2}S(1s, 2p)$. Since $\theta < 90°$, the overlap in the e orbitals is less for the pyramidal molecule. Therefore, the planar structure is favored by the greater overlap in the $1e'$ orbitals, but the pyramidal structure is favored by the s-p_z mixing in the $2a_1$ and $3a_1$ orbitals. These features are represented in the correlation diagram given in Fig. 5.4 (the $4a_1$ orbital will probably be less stable than the $3a'_1$, as indicated in the diagram). Figure 5.4 illustrates that, in fact,

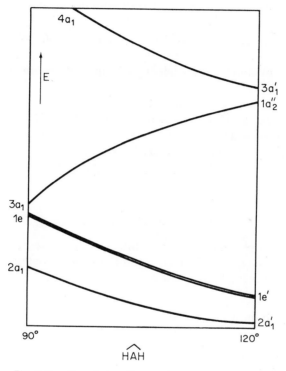

Fig. 5.4   Correlation diagram for an AH_3 molecule.

there is little to choose between the planar and pyramidal structures for NH_3. However, proton repulsion energies, discussed in Chapter 4, need to be considered. The energy has been evaluated by using the expression

$$E_{nbr} = \frac{3}{2}\left(\frac{AIS_n^2}{1 - S_n^2}\right) \tag{5.15}$$

where $A = 0.65$, I is the ionization potential of a hydrogen atom, and S_n is the overlap integral. For various bond angles, the repulsion energy is as follows:

Angle	H···H distance	S_n	E_{nbr}(eV)
120°	176	0.29	1.21
106° 46′	162.8	0.33	1.66
90°	143.7	0.41	2.65

Hence, the actual shape of the NH_3 molecule is a balance among three factors: (1) the bonding energy of the a_1 orbitals; (2) the overlap of the e orbitals; and (3) the repulsion energy of the hydrogen atoms. The net result is a small stabilization of 0.26 eV for an angle of 106° 46′ over a planar structure. This suggests that the improved bonding owing to the s-p_z mixing in the pyramidal molecule is a dominant feature of the stereochemistry.

It is now instructive to compare these results with a similar investigation for phosphine. The HPH bond angle of PH_3 is 93° 18′, and the stabilization over a planar structure is now 1.3 eV. The electron configurations for both a planar and pyramidal structure for PH_3 are

$$(1a_1')^2(2a_1')^2(1e')^4(1a_2'')^2(3a_1')_b^2(2e')_b^4(2a_2'')_{nb}^2$$

and   (5.16)

$$(1a_1)^2(2a_1)^2(1e)^4(3a_1)^2(4a_1)_b^2(2e)_b^4(5a_1)_{nb}^2$$

respectively. Phosphorus differs from nitrogen in one respect: it has outer $3d$ orbitals that may, under certain conditions, participate in the bonding. The symmetry d orbitals that can mix with the s and p orbitals are listed

Table 5.3   Symmetry Properties of d Orbitals for Point Groups C_{3v} and D_{3h}

C_{3v}		D_{3h}	
d orbital		d orbital	
d_{z^2}	a_1	d_{z^2}	a_1'
$d_{x^2-y^2}, d_{xy}$	e	$d_{x^2-y^2}, d_{xy}$	e'
$d_{xz} d_{yz}$	e		

in Table 5.3. In the pyramidal structure, there are two more d orbitals for bonding than in the planar structure. Therefore, both the s-$p_z(a_1)$ mixing and the $3d_{xz}$, $3d_{yx}(e)$ mixing favor the pyramidal structure. In addition, the nonbonded atom and bonding-nonbonding electron pair repulsions are less for PH_3 than for NH_3, owing to the larger covalent radius of phosphorus.

We can conclude that the HPH bond angle of PH_3 should be less than the bond angle of NH_3, which is what is observed. A factor ignored in the preceding discussion is the electronegativity of the nitrogen and phosphorus. The lower electronegativity of the phosphorus also aids in the formation of a smaller HPH bond angle compared with the HNH angle in ammonia.

We could argue that the size of the phosphorus atom compared with nitrogen (the covalent radii are 110 and 70 pm, respectively) is the main factor contributing to the weaker repulsions, and therefore the smaller bond angle, rather than consider d orbitals to be involved in the bonding.

The correlation diagram given in Fig. 5.4 has a further use in relating molecular shape to electronic configuration. Consider the following series of hydrides:

Hydride	BH_3	CH_3^+	$\cdot CH_3$	CH_3^-	NH_3	OH_3^+
Total number of electrons	8	8	9	10	10	10
Shape	trigonal	trigonal	planar	$\longleftarrow$ pyramidal $\longrightarrow$		

For the species with up to 8 electrons (6 valence electrons), the maximum bonding energy is achieved with planar stereochemistry, but for 10 and 11 electrons (8, 9 valence electrons), maximum bonding occurs for the pyramidal structure. The methyl radical $\cdot CH_3$ has seven electrons in the valence shell and may be expected to have a pyramidal shape. Experimental evidence from esr studies indicate, however, that it is planar. If the antibonding orbitals are also included in the correlation diagram, we can see that for a 12 or more electron system (10 or more valence electrons) the molecule should be planar. However, no such example exists for a hydride. It is simpler just to consider the valence electrons, i.e., those in the quantum levels involved in the bonding, and it is usual, in correlation diagrams such as Fig. 5.4, to exclude the $1a$ and $1a_1'$ orbitals involving atomic orbitals below the valence shell. A number of these diagrams determined by Walsh[6] for other molecular types are given in Figs. 5.5 through 5.8, and Table 5.4 summarizes the results. The diagrams can also be used in predicting the shape of excited molecules; e.g., the excited state for ammonia should be planar.

In the approach described, the influence of lone pairs on the stereochemistry is not as obvious as in the VB theory. However, the lone pairs are important. For example, the compounds ClF_3 and BrF_3 are of the AB_3 type and have 28 valence electrons; according to the correlation diagram (Fig. 5.8), they would be expected to be trigonal planar in shape. The compounds are planar but the bond angles are not 120°, and in this case the lone pairs are influencing the structure and need to be considered.

Table 5.4   Shapes of Molecules According to Walsh[6]

Compound type	Stereochemistry	Valence electrons	Examples
AH_2	Linear	4	HgH_2, BeH_2
	Bent	5–8	H_2O, H_2S
HAB	Linear	10 or less, 15–16	HCN
	Bent	11–14	HOCl
AB_2	Linear	16, 22	SCN^-, CO_2, XeF_2, I_3^-
BAC	Bent	17–20	NO_2, SO_2, Cl_2O
AH_3	Planar	6, 10	BH_3, CH_3^+
	Pyramidal	7–9	NH_3, PH_3
H_2AB	Planar	12	H_2CO, H_2BF
	Pyramidal	14	H_2NF
AB_3	Planar	24	$COCl_2$, $GaCl_3$, BF_3
B_2AC	Pyramidal	25, 26	IO_3^-, PX_3, $SOCl_2$
$HAAH$	Linear	10	C_2H_2
	Bent but planar	11, 12	N_2F_2, H_2N_2
	Bent nonplanar	14	$H_2O_2^-$

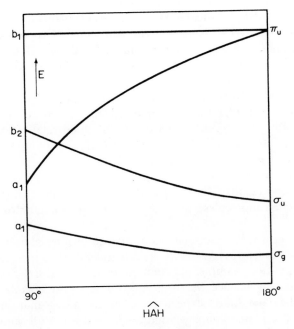

Fig. 5.5   Correlation diagram for an AH_2 molecule.

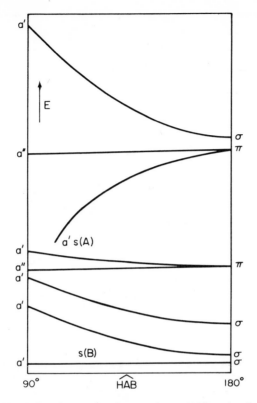

Fig. 5.6   Correlation diagram for an HAB molecule.

5.4   MULTIPLE BONDS

An important fact that has been ignored so far is multiple bonding. For example, in the planar structure of AB_3 molecules, the p_z orbital of A is nonbonding and has π symmetry. It can be used in π bonding with π orbitals of the ligands B. Hence, if π bonding is possible, it will favor a planar structure, as occurs in the species, BX_3, (X = halogen), NO_3^-, CO_3^{2-}, and SO_3.

An excellent example of the influence of π bonding on stereochemistry occurs for the compound $N(SiH_3)_3$, which is trigonal planar and not pyramidal, as perhaps expected. In the planar configuration, the filled non-σ-bonding $2p_z$ orbital perpendicular to the Si_3N plane can overlap with empty $3d$ orbitals of the silicon forming a $p_\pi \rightarrow d_\pi$ bond; such a bond cannot be formed for a pyramidal molecule. The shape of $N(CH_3)_3$, on the other hand, is pyramidal because the $p_\pi \rightarrow d_\pi$ bond is not possible inasmuch

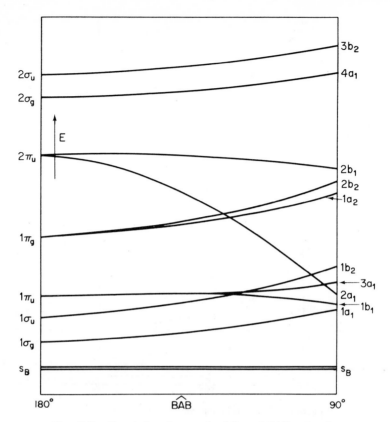

Fig. 5.7   Correlation diagram for AB_2 and BAC molecules.

as the carbon atom does not have accessible d orbitals. The weak basicity of $N(SiH_3)_3$ compared with $N(CH_3)_3$ also indicates that the nitrogen "lone pair" is bonding in the former compound.

We shall now consider two alternative approaches[9] to the influence of multiple bonding in stereochemistry that apparently give different answers. For the molecule ethylene, the σ bonding can be considered as involving sp^2 hybrids of the carbon atom and the π bond as overlap of the half-filled unused p orbitals on each carbon atom (Fig. 5.9). It is predicted, therefore, that the HCH and HCC bond angles should be close to 120°.

An alternative approach is to consider the orbital hybridization around each carbon as sp^3. The C—C bond now consists in two bent σ bonds (Fig. 5.10). In terms of this description, the bond angles HCH and HCC are 109° 28′ and 125° 16′, respectively (Table 5.5).

The bond angles for a number of similar compounds are given in Table 5.6, and it is clear that the results do not distinguish between the two

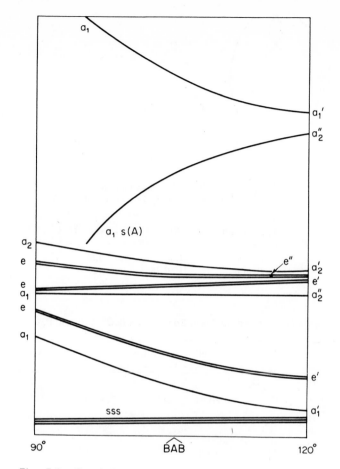

Fig. 5.8 Correlation diagrams for AB_3 and B_2AC molecules.

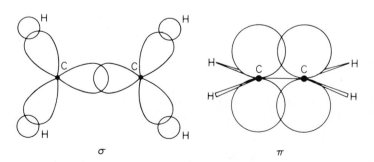

Fig. 5.9 Ethylene: σ and π orbitals.

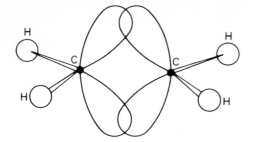

Fig. 5.10   Bent σ bonds in ethylene.

Table 5.5   Predicted Bond Angles in
Compounds $R_2C{=}CR_2$ and $R_2C{=}X$

C—C Bond	$\widehat{RCR}$	$\widehat{RCC}$ or $\widehat{RCX}$
$\sigma + \pi$	120°	120°
Bent σ	109° 28′	125° 16′

Table 5.6   Bond Angles in Compounds $R_2C{=}CR_2'$ and $R_2C{=}X$

Compound	Angle	Value	Angle	Value
$\begin{array}{c} H \\ \diagdown \\ \end{array} C{=}C \begin{array}{c} H \\ \diagup \\ \end{array}$ $\begin{array}{c} H \end{array} \quad \begin{array}{c} H \end{array}$	HCH	117°	HC=C	121.5°
$\begin{array}{c} H \\ \diagdown \\ H \diagup \end{array} C{=}O$	HCH	$118 \pm 2°$	HC=O	119.2°
$\begin{array}{c} Cl \quad\quad H \\ \diagdown \quad \diagup \\ C{=}C \\ \diagup \quad \diagdown \\ Cl \quad\quad H \end{array}$	ClCCl	122°	ClC=C	123.2°
$\begin{array}{c} CH_3 \\ \diagdown \\ CH_3 \diagup \end{array} C{=}O$			CC=O	118.5°
$\begin{array}{c} F \\ \diagdown \\ F \diagup \end{array} C{=}O$	FCF	$112.5 \pm 6°$		
$\begin{array}{c} Cl \\ \diagdown \\ Cl \diagup \end{array} C{=}O$	ClCCl	111.3°		

Table 5.6 *(cont.)*

Compound	Angle	Value	Angle	Value
CH_3, CH_3 / C=C \ H, H	CCC	114°	HC=C	109.5°
F, F / C=C \ F, Cl	FCF	114°	FC=C	123°
Cl, Cl / C=C \ Cl, Cl	ClCCl	113° 18′		
F, F / C=C \ F, F	FCF	110°		
F, H / C=C \ F, H	HCH	120°		
	FCF	109° 3′		

descriptions. Nonbonded atom repulsions will affect the bond angles, and it is possible that these interactions dominate.[10] A number of other bond angles could be cited for systems such as $RR'C=C$ and $RR'C=X$, but in these cases the difference in the groups R and R', such as electronegativity and bond radius, introduces further unknowns into the estimate of bond angles. The inability of resolving what appears to be a straightforward problem underlines the difficulty we have in discussing the finer details of stereochemistry.

BIBLIOGRAPHY

1. J. D. ROBERTS, *Molecular Orbital Calculations*, W. A. Benjamin, Inc., New York, 1962.

2. H. B. GRAY, *Electrons and Chemical Bonding*, W. A. Benjamin, Inc., New York, 1964.

3. C. J. BALLHAUSEN and H. B. GRAY, *Molecular Orbital Theory*, W. A. Benjamin, Inc., New York, 1965.

4. F. O. ELLISON and H. SHULL, *J. Chem. Phys.*, **23**, 2348 (1955).

5. I. N. LEVINE, *Quantum Chemistry*, Vol. 2, Allyn and Bacon, Inc., Boston, 1970, p. 42 and references therein.

6. A. D. WALSH, *J. Chem. Soc.*, 2260, 2266, 2289 (1953).

7. C. A. COULSON, *Valence*, 2nd ed., Oxford University Press, London, 1961.

8. R. S. MULLIKEN, *J. Am. Chem. Soc.*, **77**, 887, (1955).

9. L. PAULING, *Kekulé Symposium* London, 1958; and *Nature of the Chemical Bond*, 3rd ed., Oxford University Press, London, 1960.

10. L. S. BARTELL, *J. Chem. Educ.*, **45**, 754 (1968).

FURTHER READING

W. J. ORVILLE-THOMAS, *The Structure of Small Molecules*, Elsevier Publishing Co., Amsterdam, 1966.

D. J. ROYÉR, *Bonding Theory*, McGraw-Hill Book Company, New York, 1968.

J. BARRETT, *Atomic and Molecular Structure*, John Wiley & Sons, Inc., New York, 1970.

6
Equivalent Orbitals

6.1 INTRODUCTION

Application of the molecular orbital theory to stereochemical problems requires a knowledge of the point group of the molecule. Also, the delocalized molecular orbitals constructed are not very informative as to the details of the molecular shape, and only after lengthy calculations can bond angles be found. A qualitative approach to this problem is to apply the method of equivalent orbitals (EO).[1-6] Briefly, the method amounts to taking a linear combination of delocalized MO's to give a new set of equivalent directional orbitals. The equivalent orbitals are therefore similar to the VB localized bonding orbitals formed from hybridized atomic orbitals. The equivalent orbital approach is, in fact, a link between the MO and VB theories.

6.2 A SIMPLIFIED THEORY OF EQUIVALENT ORBITALS[6]

If we consider an electron restricted to movement along a wire of length l with a constant and zero potential energy, then the equation of motion is

97

given by the Schrödinger equation

$$-\frac{h^2}{8\pi^2 m}\frac{d^2\psi}{dx^2} = E\psi \tag{6.1}$$

which has a general solution

$$\psi = A \sin\left(\sqrt{\frac{8\pi^2 mE}{h^2}}x\right) + B \cos\left(\sqrt{\frac{8\pi^2 mE}{h^2}}x\right) \tag{6.2}$$

The wave function ψ must be zero at $x = 0$ and $x = l$, which means that $B = 0$ and $\sqrt{8\pi^2 mE/h^2}\,l$ is an integral multiple of π. The two orbitals of lowest energy (Fig. 6.1) are, therefore,

$$\psi_1 = \sqrt{\frac{2}{l}}\sin\frac{\pi x}{l}$$

$$\psi_2 = \sqrt{\frac{2}{l}}\sin\frac{2\pi x}{l} \tag{6.3}$$

Consider one electron (spin α) placed in each of the orbitals [Equation (6.3)]. The complete antisymmetric wave function is

$$\Psi = \frac{1}{\sqrt{2}}[\psi_1(1)\psi_2(2) - \psi_1(2)\psi_2(1)]\alpha(1)\alpha(2) \tag{6.4}$$

which in determinant form is

$$\Psi = \frac{1}{\sqrt{2}}\begin{vmatrix}\psi_1(1) & \psi_1(2) \\ \psi_2(1) & \psi_2(2)\end{vmatrix}\alpha(1)\alpha(2) \tag{6.5}$$

We can obtain equivalent orbitals from the functions ψ_1 and ψ_2 by taking the following linear combinations:

$$\chi_a = \frac{1}{\sqrt{2}}(\psi_1 + \psi_2)$$

$$\chi_b = \frac{1}{\sqrt{2}}(\psi_1 - \psi_2) \tag{6.6}$$

provided the energy difference between ψ_1 and ψ_2 is small. Both of these functions [Equation (6.6)] have a pronounced directional property, as Fig. 6.1 indicates. The complete wave function in terms of the equivalent orbitals χ_a and χ_b with an electron of spin α in each is

$$\psi_{EO} = -\frac{1}{\sqrt{2}}\begin{vmatrix}\chi_a(1) & \chi_a(2) \\ \chi_b(1) & \chi_b(2)\end{vmatrix}\alpha(1)\alpha(2) \tag{6.7}$$

Substitute for χ_a and χ_b in (6.7), according to Equation (6.6), and expand.

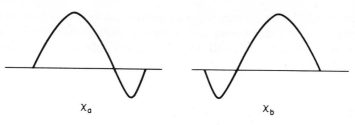

Fig. 6.1 Equivalent orbitals.

The expansion is identical to Equation (6.4), indicating that the two functions (6.5) and (6.7) are the same. This demonstrates that the delocalized wave functions ψ_1 and ψ_2 and the localized EO wave functions χ_a and χ_b are both equivalent descriptions of the same system. The formation of equivalent orbitals is, in fact, a particular case of a more general property of solutions of the Schrödinger equation—namely, a system represented by a set of wave functions $|\psi_1 \psi_2 \psi_3 \ldots \psi_n|$ has an equally valid representation in the set $|\psi_1 + \psi_2 \psi_2 \psi_3 \ldots \psi_n|$.

A distinct advantage of equivalent orbitals is that the exchange energy is reduced to a small value of the total energy. Whereas the exchange energy accounts for 28% of the total energy in the MO theory, it is only 2 to 3% of the total in the EO theory.

6.3 SYMMETRY AND EQUIVALENT ORBITALS

The equivalent orbitals constructed from the molecular orbitals are not symmetry orbitals in the way that molecular orbitals are, but they do transform into each other by the operations of the point group of the molecule under consideration. This only holds when the number of symmetry molecular orbitals combined to give equivalent orbitals does not exceed two. For example, the two symmetry orbitals a_1 and e of the C_{3v} point group can be combined to give three equivalent orbitals at an angle of $120°$ to each other, and these three orbitals can be transformed into each other by the operations of the C_{3v} point group. Likewise, the two symmetry orbitals a_1 and t_2 of the tetrahedral point group can be combined to give four equivalent orbitals at $109.5°$ to each other. On the other hand, the three symmetry orbitals $a_{1g} t_{2g} t_{1u}$ of the octahedral point group will give two sets of equivalent orbitals that are not symmetry related to each other. One of the orbitals will be at $90°$ to four others and at $180°$ to the fifth orbital. Hence, it is not possible in an octahedral configuration to write down six equivalent orbitals but only two sets—one of four and one of two equivalent orbitals. This result immediately demonstrates that the equatorial bonds

and polar bonds in the shapes trigonal bipyramid, octahedron, and pentagonal bipyramid can be different. As we have seen, this difference is observed
for a number of compounds with trigonal bipyramidal stereochemistry (see
Section 4.5).

6.4 EQUIVALENT ORBITALS FOR DIATOMIC MOLECULES

The MO description of the diatomic molecule N_2 is $(\sigma 1s)^2(\sigma^*1s)^2(\sigma 2s)^2$
$(\sigma^*2s)^2(\sigma 2p)^2(\pi 2p)^4$ (p. 81), indicating three bonding orbitals, $\sigma^2\pi^4$. These
orbitals written out in full are

$$\psi^b_{\sigma 2p} = \frac{1}{\sqrt{2}}(2p\sigma_A + 2p\sigma_B)$$

$$\psi^b_{\pi 2p_x} = \frac{1}{\sqrt{2}}(2p_x\pi_A + 2p_x\pi_B) \qquad (6.8)$$

$$\psi^b_{\pi 2p_y} = \frac{1}{\sqrt{2}}(2p_y\pi_A + 2p_y\pi_B)$$

We obtain equivalent orbitals from these three orbitals by ensuring that
they conform to a trigonal transformation. The result is

$$\chi^b_1 = \frac{1}{\sqrt{3}}\psi^b_{\sigma 2p} + \sqrt{\frac{2}{3}}\psi^b_{\pi 2p_x}$$

$$\chi^b_2 = \frac{1}{\sqrt{3}}\psi^b_{\sigma 2p} - \frac{1}{\sqrt{6}}\psi^b_{\pi 2p_x} + \frac{1}{\sqrt{2}}\psi^b_{\pi 2p_y} \qquad (6.9)$$

$$\chi^b_3 = \frac{1}{\sqrt{3}}\psi^b_{\sigma 2p} - \frac{1}{\sqrt{6}}\psi^b_{\pi 2p_x} - \frac{1}{\sqrt{2}}\psi^b_{\pi 2p_y}$$

The three new orbitals are sometimes called "bent orbitals," and have the
shape given in Fig. 6.2. The same approach can be used for double and
triple bonds in polyatomic compounds, such as the $\sigma^2\pi^2$C—C bonds in
ethylene and the $\sigma^2\pi^4$C—C bonds in acetylene (see Section 5.4).

6.5 EQUIVALENT ORBITALS FOR POLYATOMIC MOLECULES

We shall consider an example of a polyatomic molecule, which brings out
further points of the EO approach. The bonding molecular orbitals of

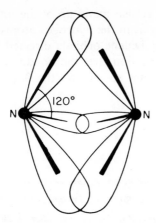

Fig. 6.2 Equivalent orbitals in N_2.

ammonia are $2a_1$ and $1e$ (p. 82). These can be combined in the form

$$\chi_n = a_n(2a_1) + b_n(1e_x) + c_n(1e_y) \tag{6.10}$$

where $n = 1, 2, 3$ and $1e_x$, $1e_y$ are the component wave functions of the doubly degenerate $1e$ function. The mathematical form of the linear combinations is similar to that of trigonal hybrids. Therefore, one would expect the coefficients a_n, b_n, and c_n to be the same as those for trigonal hybrids. The three equivalent orbitals are, therefore,

$$\chi_1^b = \frac{1}{\sqrt{3}}(2a_1) + \sqrt{\frac{2}{3}}(1e_x)$$

$$\chi_2^b = \frac{1}{\sqrt{3}}(2a_1) - \frac{1}{\sqrt{6}}(1e_x) + \frac{1}{\sqrt{2}}(1e_y) \tag{6.11}$$

$$\chi_3^b = \frac{1}{\sqrt{3}}(2a_1) - \frac{1}{\sqrt{6}}(1e_x) - \frac{1}{\sqrt{2}}(1e_y)$$

The trigonal form of Equations (6.11) appears to predict that the NH_3 molecule is planar with a bond angle of $120°$.[7] However, if one looks closely at the composition of the $2a_1$ function (p. 85), it is clearly a mixture of the $2s$ and $2p_z$ atomic orbitals of the nitrogen atom. This means that the bonding orbital $2a_1$ is not spherically symmetrical but has its greatest electron density on the side of the nitrogen closest to the three hydrogens. In other words, although we have three apparently trigonal equivalent orbitals the s-p mixing reduces the HNH bond angle from $120°$, but not to less than $90°$. The actual value depends on the relative proportion of the $2s$ to $2p_z$ functions in the $2a_1$ orbital.

The nonbonding $3a_1$ orbital is also made up of the nitrogen $2s$ and $2p_z$ orbital, this time pointing away from the hydrogen atoms. This orbital could be included to give four equivalent orbitals directed tetrahedrally. However, this would give the impression that the bonding and lone pair orbitals were equivalent, which is not so. Hence, the $3a_1$ orbital is best excluded in the EO approach.

BIBLIOGRAPHY

1. J. E. LENNARD-JONES, *Proc. Roy. Soc., London, Ser. A*, **198**, 14 (1949).

2. G. G. HALL and J. E. LENNARD-JONES, *Proc. Roy. Soc., London, Ser. A*, **202**, 155 (1950).

3. J. E. LENNARD-JONES and J. A. POPLE, *Proc. Roy. Soc., London, Ser. A*, **202**, 166 (1950).

4. G. G. HALL, *Proc. Roy. Soc., London, Ser. A*, **202**, 336 (1950).

5. J. E. LENNARD-JONES, *Proc. Roy. Soc., London, Ser. A*, **198**, 1 (1949).

6. J. A. POPLE, *Quart. Rev. London*, **11**, 273 (1957).

7. J. C. SLATER, *Quantum Theory of Molecules and Solids*, Vol. 1, McGraw-Hill Book Company, New York, 1963.

7
Transition Metals

7.1 INTRODUCTION

The transition metals have a penultimate d shell that may be partly filled with electrons. The availability of the $(n - 1)d$ orbitals for σ bonding depends on whether they are occupied by the d electrons. The stereochemistry of transition metal complexes depends on two main factors: (1) the number and type of d orbitals used in bonding; (2) the arrangement of the metal d electrons in the remaining $(n - 1)d$ orbitals. We shall discuss these aspects by considering first the effect of different stereochemical arrangements of ligands on the metal d orbitals, and second the effect of the metal d electrons on the ligands.

7.2 STEREOCHEMISTRY AND d ORBITALS

One of the most common stereochemistries for transition metal compounds is the octahedron; therefore, we shall consider a regular octahedral complex ML_6, where M is a first-row transition metal (i.e., $3d$ orbitals involved). The bonding will be described in terms of Fig. 7.1.[1,2]

103

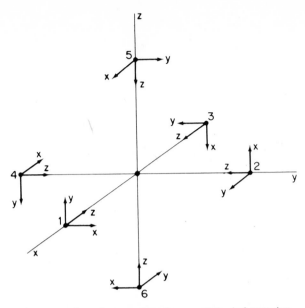

Fig. 7.1   Coordinate system for an octahedral complex.

The atomic orbitals on the metal that can be used in bonding are the $3d$, $4s$, and $4p$. In octahedral symmetry (O_h), the symmetry orbitals are (Table 2.11)

$$3d \quad e_g \qquad d_{x^2-y^2}, d_{z^2}$$
$$\qquad t_{2g} \qquad d_{xy}, d_{yz}, d_{zx}$$
$$4s \quad a_{1g}$$
$$4p \quad t_{1u}$$

The first obvious fact from a consideration of symmetry is that the five d orbitals fall into two symmetry classes. The next step is deciding which of the nine orbitals are used in the six σ bonds. To do this, we must find the transformation properties of the six M—L σ bonds in the O_h point group. The trace of each transformation matrix is listed under the appropriate symmetry in Table 7.1. The reducible representation Γ_σ reduces to $\Gamma_\sigma =$

Table 7.1   Reducible Representation of the σ Bonds in an Octahedral Complex

E	$8C_3$	$3C_2$	$6C_4$	$6C_2'$	i	$8iC_3$	$3iC_2$	$6iC_4$	$6iC_2'$
6	0	2	2	0	0	0	4	0	2

$a_{1g} + t_{1u} + e_g$. Therefore, only two d orbitals are used in the σ bonds in an octahedral complex: $d_{x^2-y^2}$ and d_{z^2}. The next step is to determine the form of the six σ-bonding molecular orbitals of the six ligands that transform as a_{1g}, e_g, and t_{1u}. Use the method described on p. 83. The results for an octahedral complex are given in Table 7.2.

Table 7.2 Molecular Orbital Combinations in O_h Symmetry

Symmetry	Central atom	Ligand orbitals	Ligand orbitals
a_{1g}	s	$\dfrac{1}{\sqrt{6}}(z_1 + z_2 + z_3 + z_4 + z_5 + z_6)$	
e_g	$d_{x^2-y^2}$	$\dfrac{1}{2}(z_1 - z_2 + z_4 - z_5)$	
	d_{z^2}	$\dfrac{1}{3\sqrt{2}}(2z_3 + 2z_6 - z_1 - z_2 - z_4 - z_5)$	
t_{1u}	p_x	$\dfrac{1}{\sqrt{2}}(z_1 - z_4)$	$\dfrac{1}{2}(x_3 + y_2 - x_5 - y_6)$
	p_y	$\dfrac{1}{\sqrt{2}}(z_2 - z_5)$	$\dfrac{1}{2}(x_1 + y_3 - x_6 - y_4)$
	p_z	$\dfrac{1}{\sqrt{2}}(z_3 - z_6)$	$\dfrac{1}{2}(x_2 + y_1 - x_4 - y_5)$
t_{2g}	d_{xz}		$\dfrac{1}{2}(x_3 + y_1 + x_4 + y_6)$
	d_{yz}		$\dfrac{1}{2}(x_2 + y_3 + x_6 + y_5)$
	d_{xy}		$\dfrac{1}{2}(x_1 + y_2 + x_5 + y_4)$

We can now write the six molecular orbitals of the octahedral complex. They have the general form

$$\psi = a\psi_{\text{metal}} + b\psi_{\text{ligand}} \tag{7.1}$$

For example,

$$\psi_{e_g}^I = ad_{x^2-y^2} + b\tfrac{1}{2}(z_1 - z_2 + z_4 - z_5) \tag{7.2}$$

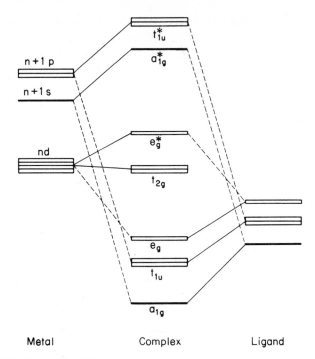

Fig. 7.2   Simplified molecular orbital diagram for an octahedral complex ML_6, only σ bonding is considered.

A qualitative MO energy diagram for the octahedral complex is given in Fig. 7.2.

The metal d orbitals are split into two groups, and the energy of separation $t_{2g} - e_g^*$ is designated $10Dq$ or Δ. Dq is a parameter and is proportional to $\bar{a}^4/r^5$ or $\bar{a}^4/r^6$, where $\bar{a}^4$ is the average fourth-power value of the radius of the d electrons and r is the M—L bond length.

The t_{2g} orbitals have π symmetry with respect to the six ligands. They are capable of forming either M → L or M ← L π bonds with either empty or filled π orbitals of the ligands, respectively. The ligand π orbitals are also given in Table 7.2. The effect of bonding on the energy $10Dq$ is to either increase it or decrease it, depending on whether the bond is M → L or M ← L.

Various approximations of the molecular orbital approach, such as the crystal field and ligand field theories, have been used with success in explaining the properties of transition metal complexes. These methods are described in detail in various texts (see Bibliography and Further Reading).

The magnitude of $10Dq$ ranges from 5000 to 40,000 cm^{-1}, depending on the stereochemistry, coordination number, and charge on the central metal ion. For octahedral stereochemistry, $10Dq \sim 10,000 \text{ cm}^{-1}$ for divalent cations and $\sim 20,000 \text{ cm}^{-1}$ for trivalent cations for the first transition metal series. For tetrahedral stereochemistry, $10Dq$ is approximately four-ninths of the preceding figures.

The splitting of the d orbitals for other stereochemistries can be obtained in a similar manner. Purely on symmetry grounds, we can see how the orbitals split. For example, for a square pyramidal complex, with point group symmetry C_{4v}, the d orbitals split (Table 2.11) as follows:

$$
\begin{array}{cc}
d_{z^2} & a_1 \\
d_{x^2-y^2} & b_1 \\
d_{xy} & b_2 \\
d_{yz}\, d_{zx} & e
\end{array}
$$

However, the relative order of energies of the orbitals is not found so readily. But, by considering which of the orbitals are bonded (or nearest) to the ligands, we can obtain a relative order. For a square pyramid, as in

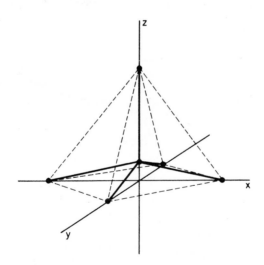

Fig. 7.3   Coordinate system for a square pyramid.

Fig. 7.3, it is clear that the $d_{x^2-y^2}$ and d_{z^2} orbitals are the bonding orbitals— $d_{x^2-y^2}$ more than the d_{z^2}, because its four lobes bond to four ligands whereas only one lobe of the d_{z^2} orbital is used. Of the remaining orbitals, the d_{xy} has more ligands nearer it than either the d_{xz} or d_{yz} orbitals. Hence, it will be destabilized more than the latter two, but not as much as the antibonding states of the $d_{x^2-y^2}$ and d_{z^2} orbitals. Therefore, a relative ordering of the metal orbitals in decreasing stability is

$$(d_{xz}d_{yz}), d_{xy}, d^*_{z^2}, d^*_{x^2-y^2}$$

The results of crystal field calculations of d-orbital energies for a one-electron system are given in Table 7.3.[3]

Table 7.3 d-Orbital Energies in Various Stereochemistries (*Dq* Units)

CN	Shape	$d_{x^2-y^2}$	d_{z^2}	d_{xy}	d_{xz}	d_{yz}	Comment
1		-3.14	5.14	-3.14	0.57	0.57 ⎱	Bond along
2	Linear	-6.28	10.28	-6.28	1.14	1.14 ⎰	z axis
3	Trigonal plane	5.46	-3.21	5.46	-3.86	-3.86	Bonds in xy plane
4	Tetrahedral	-2.67	-2.67	1.78	1.78	1.78	
4	Square plane	12.28	-4.28	2.28	-5.14	-5.14	Bonds in xy plane
5	Trigonal bipyramid	-0.82	7.07	-0.82	-2.72	-2.72 ⎱	Pyramid base
5	Square pyramid	9.14	0.86	-0.86	-4.57	-4.57 ⎰	in xy plane
6	Octahedron	6.00	6.00	-4.00	-4.00	-4.00	
7	Pentagonal bipyramid	2.82	4.93	2.82	-5.28	-5.28	Pyramid base in xy plane
7	Distorted octahedron	8.79	1.39	-1.51	-2.60	-6.08	
			(hybrids)		(hybrids)		
8	Dodecahedron, square antiprism	Order and values vary markedly with bond angles.					

7.3   *d*-ELECTRON ARRANGEMENT

So far, we have considered the effect of different ligand arrangements on the d orbitals, assuming that all the ligands are of the same type and that the model is mainly electrostatic in nature. We next consider the factors that may distort the stereochemistry, the principal of which is the arrangement of the metal d electrons. We shall consider in detail the octahedral stereochemistry, and conclude with a brief survey of the less common stereochemistries.

If we consider the MO diagram, Fig. 7.2, the six lowest lying orbitals, e_g, t_{1u}, a_{1g}, are filled with 12 ligand σ electrons. The metal d electrons will start to fill the t_{2g} orbitals (t^b_{2g} when a M $\rightarrow$ L π bond occurs, or t^*_{2g} when a M $\leftarrow$ L π bond occurs). The magnitude of $10Dq$ is comparable with that of the interelectronic repulsion energy (R.E.) Therefore, either the t_{2g} orbitals will fill singly and then e_g^* if $10Dq <$ R.E., or electrons will pair up in the t_{2g} orbitals before going into the e_g^* orbitals if $10Dq >$ R.E. Since $10Dq$ varies with the type of bonding in a complex and with the oxidation state of the metal, and also since R.E. varies with the size of the d orbitals, it being less for the second- and third-row transition metal series owing to the more diffuse d orbitals allowing the electrons to get further apart, one finds that spin pairing is more common for metal ions in high oxidation states, and complexes formed with π-bonding ligands. Also, spin pairing always occurs for the heavy transition metals. The arrangement of electrons in the d orbitals for other stereochemistries can be considered similarly, i.e., spin pairing being dependent on the magnitude of the orbital splitting.

7.4 INFLUENCE OF d ELECTRONS ON STEREOCHEMISTRY

The metal d electrons do not have the same type of influence on stereochemistry as lone pairs do for the regular elements. For example, in $MoF_6^{2-}[Mo(IV)d^2]$, the two unpaired d electrons do not occupy stereochemical sites giving a distorted octahedral stereochemistry. The d electrons are in t_{2g} orbitals, which are pointing between the ligands. These orbitals are not directed as, say, p orbitals or hybrid orbitals, but lie in planes. Therefore, their influence is less directional. Also, for metal ions with more than two d electrons and surrounded by six ligands, it would be virtually impossible to arrange both electrons and ligands so they *all* occupy what are effective coordination sites around the ion.

Another factor that may prevent d electrons from occupying coordination sites around an atom is their use in $M \longrightarrow L$ π bonding, especially for the platinum metals and metal ions in low oxidation states. Also, in a situation such as MoF_6^-, if the single d electron occupied a face of the octahedron and hence distorted it, a consequence would be to alter the ligand field splitting of the d orbitals—a situation energetically less favorable for the complex.

Nevertheless, d electrons can have an influence on transition metal stereochemistry by virtue of a different mechanism from that of lone pairs. The charge field produced by the electrons in the d orbitals on the ligands around the central atom can be either spherically symmetric or unsymmetric. In the first case, the stereochemistry is not affected; however, if the field is not symmetric, i.e., the effect on each ligand is not the same, it can alter the shape of the molecule. The results of the application of this principle to the configurations d^0 through d^{10} are given in Table 7.4.[4]

Consider the case of a d^4 spinfree configuration in an octahedral complex. The ground term ^5E_g can arise in one of two ways: either

Table 7.4 d-Electron Configuration and Shape of Four- and Six-Coordinate Complexes

Number of nonbonding d electrons	Unpaired electrons	Four-coordinate	Six-coordinate
		High spin	
0, 10, or 5	0 or 5	Regular tetrahedron	Regular octahedron
9 or 4	1 or 4	Square planar	Tetragonal
8 or 3	2 or 3	Distorted tetrahedron	Regular octahedron
7 or 2	3 or 2	Regular tetrahedron	Almost regular octahedron
6 or 1	4 or 1	Almost regular tetrahedron	Almost regular octahedron
		Low spin	
3	1	Almost regular tetrahedron	
4	0 or 2	Regular tetrahedron	Almost regular octahedron
5	1	Distorted tetrahedron	Almost regular octahedron
6	2 or 0	Distorted tetrahedron	Regular octahedron
7	1	Square plane	Tetragonal
8	0	Square plane	Tetragonal

$(d_{x^2-y^2})^1(d_{z^2})^0$ or $(d_{z^2})^1(d_{x^2-y^2})^0$. In this situation, it is usual, but not always correct, to invoke the Jahn-Teller theorem,* which predicts that the complex will distort so as to remove the degeneracy of the e_g orbitals. The resultant orbital energy diagram given in Fig. 7.4 is the result of a tetragonal distortion (elongation of the complex in the z direction). For the t_{2g} orbitals, the net overall energy change is zero,

$$2 \times \tfrac{1}{3}(-\delta_2) + 1 \times \tfrac{2}{3}(\delta_2) = 0$$

whereas for the e_g orbitals it is $(-\tfrac{1}{2}\delta_1)$. This additional stabilization is the driving force of the distortion. Manganese in MnF_3 has a distorted octahedral stereochemistry; the Mn—F bond lengths are 179, 191, and 209 pm. Unfortunately, we cannot decide which of the e_g orbitals is stabilized, for either will give the same net stabilization energy. Experimental results indicate that it is usually the d_{z^2} orbital, although some examples of stabilization of the $d_{x^2-y^2}$ orbital are known. The Jahn-Teller distortion is the reason

* The Jahn-Teller theorem states that any nonlinear molecular system having a degenerate electronic state will distort to lower the molecular symmetry and destroy the degeneracy.

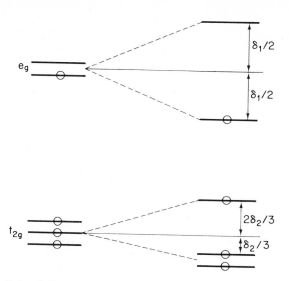

Fig. 7.4 Splitting of orbital degeneracy for an octahedral complex with the d^4 spinfree configuration.

normally given for the large number of distorted octahedral Cu(II) complexes, where the ground term ^5E_g is orbitally degenerate.

The distortion is expected to be greatest when the orbital degeneracy occurs in the orbitals that lie in the direction of the ligands because the charge field produced by the nonbonding d electrons in these orbitals will repel the ligands. When the degeneracy occurs in the t_{2g} orbitals (octahedral complex), the distortion is much less because the t_{2g} orbitals lie between the ligands and any unsymmetric field is less effective. Distortion arising from this effect has not yet been observed structurally.

The Jahn-Teller effect is also active for excited states, and distortions may occur not present in the ground state. Take, for example, Ti(III)(d^1). The ground term in an octahedral complex is $^2T_{2g}$, and it may have a weak Jahn-Teller distortion. However, in the excited state $(e_g)^1$, the term is ^2E_g, and the complex will distort, giving the relative orbital energies as in Fig. 7.4. Hence, the single spectral absorption $T_{2g} \rightarrow E_g$ of the Ti(III) ion should be resolved into two bands. If the splitting in the T_{2g} state is ignored, the bands will correspond to the transitions $t_{2g} \rightarrow d_{z^2}$ and $t_{2g} \rightarrow d_{x^2-y^2}$. The energy difference is δ_1. The broadness of the absorption band for Ti(H$_2$O)$_6^{3+}$ is said to be caused by the described splitting. The effect is quite marked for Co(III) in CoF$_6^{3-}$ ($t_{2g}^4 e_g^2$), where the two bands are resolved (Fig. 7.5).

The approach used for octahedral stereochemistry can also be used for tetrahedral stereochemistry, which also has cubic symmetry. Table 7.4 gives the expected distortions. For stereochemistries other than octahedral or

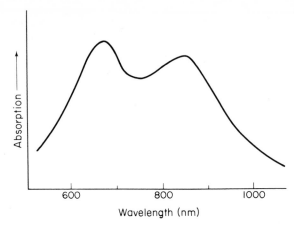

Fig. 7.5   Absorption spectrum of the $[CoF_6]^{3-}$ ion.

tetrahedral ones, orbital degeneracy is generally less common and the Jahn-Teller effect would not be expected to operate to the same extent.

It is necessary to point out that the ground state (or excited state) degeneracy can be removed by means other than the Jahn-Teller effect. For example, by spin-orbit coupling or nonequivalent ligands. Therefore, strictly, one should only consider the Jahn-Teller effect for metal atoms where the spin-orbit coupling energy is small (first-row transition metals) and where all the ligands around the metal ion are identical.

7.5   ISOMERISM

Both optical and geometric isomerism can occur in coordination compounds, and most particularly for square planar, tetrahedral and octahedral complexes.

Geometric Isomerism

Square Planar. The number of geometric isomers for square planar complexes with common stoichiometries is given in Table 7.5. Dimeric complexes with a square planar stereochemistry around the metal atoms also show cis-trans isomerism. For example, $[PtCl_2R_3P]_2$ forms the isomers

Table 7.5 Geometric Isomers for Square Planar
Complexes

Compound type	Number of geometric isomers
MA_2B_2	2 (cis and trans)
MA_2BC	2 (cis and trans)
$MABCD$	3
$M(AB)C_2^*$	2
$M(AB)CD$	2
$M(AB)_2$	2 (cis and trans)
$M(AB)(CD)$	2

* (AB) = Unsymmetric bidentate.

whereas the unsymmetric structure

is not known.

The preparation of either a cis or a trans isomer is often determined
by the nature of the ligands coordinated to the metal. For example, the
following two reactions are specific for forming *cis*- and *trans*[$PtCl_2(NH_3)_2$],
respectively:

$$[PtCl_4]^{2-} \xrightarrow{NH_3} [PtCl_3NH_3]^- \xrightarrow{NH_3} cis[PtCl_2(NH_3)_2]$$

$$[Pt(NH_3)_4]^{2+} \xrightarrow{Cl^-} [Pt(NH_3)_3Cl]^+ \xrightarrow{Cl^-} trans[PtCl_2(NH_3)_2]$$

The reason for the specificity is given in terms of the trans effect of ligands.
Certain ligands labilize ligands trans to themselves more readily than do
others, and it is possible to list ligands in an increasing order of trans effect.
In the first of the preceding reactions, the trans directing influence of Cl is
greater than NH_3, and therefore the Cl cis to the NH_3 is labilized rather
than the one trans to it. In a like manner the second reaction can be explained.
The order of ligands in increasing trans effect is $H_2O < OH^- < NH_3 \sim$
$RNH_2 <$ pyridine $< Cl^- < Br^- < NCS^- \sim I^- \sim NO_2^- \sim R^- < H^- \sim$
$SO_3H^- \sim PR_3 \sim R_2S \sim SC(NH_2)_2 < NO \sim CO \sim C_2H_4 \sim CN^-$. With
some exceptions, the order correlates reasonably well with the accepted
π-bonding properties of the ligand, and it is possible that π bonding is one
factor in the trans effect. Ligands, such as $-R^-$ and $-H^-$, also have a

strong trans effect, and these cannot participate in π bonding, indicating that σ bonding is also an important factor.

Octahedral Complexes. Geometric isomers are common for octahedral complexes; Table 7.6 gives a few examples. Methods are available for working out the numbers of isomers; they can be found in a number of sources.[3,5,6]

Table 7.6   Geometric Isomers for Octahedral Complexes

	Total isomers	Number optically active
$M(ABCDEF)$	30	30
$M(ABCDE_2)$	15	12
$M(ABC_2D_2)$	8	4
$M(A_2B_2C_2)$	6	2
$M(A_3B_3)$	2	
$M(A_4B_2)$	2	
$M(AB)CDEF^*$	24	24
$M(AA)BCDE^*$	12	12
$M(AA)B_2C_2$	4	2
$M(AB)_2CD$	11	10
$M(AA)_2BC$	3	2
$M(AA)_2B_2$	3	2

* (AB) and (AA) = Bidentate.

The geometric isomers *cis*- and *trans* $M(A_4B_2)$ and *mer*- and *fac* $M(A_3B_3)$ are common, and will be discussed in more detail. The stereochemistry for each corresponds to a slight distortion of the regular octahedral symmetry, and can be considered initially from an octahedral viewpoint. Then, by the method of "descent in symmetry," the effect of the new symmetry can be determined.[1] *Trans* $M(A_4B_2)$ has D_{4h} symmetry, and the change in degeneracy of the orbitals and the electronic energy states on passing from O_h to D_{4h} symmetry can be obtained from descent in symmetry tables.

Orbitals		Terms	
O_h	D_{4h}	O_h	D_{4h}
t_{2g}	$e_g + b_{2g}$	A_{1g}	A_{1g}
		A_{2g}	B_{1g}
e_g	$b_{1g} + a_{1g}$	E_g	$A_{1g} + B_{1g}$
		T_{1g}	$A_{2g} + E_g$
		T_{2g}	$B_{2g} + E_g$

The species $cis(MA_4B_2)$ has C_{2v} symmetry, and the change is as follows:

O_h	C_{2v}		O_h	C_{2v}
t_{2g}	$a_1 + b_1 + b_2$		A_{1g}	A_1
			A_{2g}	A_2
e_g	$a_1 + a_2$		E_g	$A_1 + A_2$
			T_{1g}	$A_2 + B_1 + B_2$
			T_{2g}	$A_1 + B_1 + B_2$

The first-order energy diagrams for these symmetries are given in Fig. 7.6.

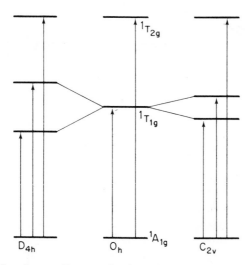

Fig. 7.6 Energy diagrams for the point group symmetries D_{4h}, O_h, and C_{2v} for the d^6 spin paired configuration.

From the diagrams, it is clear that some spectral differences between the *cis-* and *trans*MA_4B_2 would be expected, and this can be used as a method of distinguishing the isomers. The two ligand field bands $^1A_{1g} \rightarrow {}^1T_{1g}$ and $^1A_{1g} \rightarrow {}^1T_{2g}$ in a regular octahedral complex become three in the cis and trans complexes owing to the splitting of the $^1T_{1g}$ term. The splitting is greatest for the trans complex, which is reflected in the separation of band maxima in the electronic spectra. In addition, the center of symmetry is lost for the cis complex, and the band intensities should be greatest in this case. These conclusions are borne out in the spectra of *cis-* and *trans*[Co(en)$_2$F$_2$]$^+$, where, in fact, the splitting of the $^1T_{1g}$ state is not resolved in the cis complex (Fig. 7.7). The same approach may be applied to complexes MA_3B_3, where the 1, 2, 3, or fac complex has cubic symmetry and two bands are expected in, say, a Co(III) complex, whereas the 1, 2, 6, or mer complex has rhombic symmetry and six bands are expected. In fact, six bands are never seen, owing to the problem of resolution, but broadening of absorptions is common.

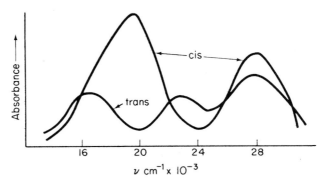

Fig. 7.7   Absorption spectra for *cis* and *trans* $[Co(en)_2F_2]^+$.

Optical Isomers

Optical isomerism is found mainly for tetrahedral and octahedral complexes, and has been used in some cases to distinguish tetrahedral from square planar coordination.

The complex types most widely studied are of the form $M(AA)_3$ and $M(AA)_2B_2$, where the enantiomorphs for the two complex types are given in Fig. 7.8. The electronic configurations for which optical isomers can be resolved most readily are t_{2g}^6 and t_{2g}^3. These configurations are associated with the most inert complexes, and racemization is not so much of a problem.

Diastereoisomerism can occur when optical activity is also present in the ligand. For example, the bidentate amine propylenediamine forms the complex ion $[Co(pn)_3]^{3+}$, and there are eight possible isomers, as follows:

$$
\begin{array}{ll}
D \; l \; l \; l & L \; l \; l \; l \\
D \; l \; l \; d & L \; l \; l \; d \\
D \; l \; d \; d & L \; l \; d \; d \\
D \; d \; d \; d & L \; d \; d \; d
\end{array}
$$

where D and L refer to the cobalt symmetry and d and l to the ligand. However, steric factors arising from ligand and metal conformations generally determine which isomer forms in the greatest yield; for example, with $l - pn$ the major product is $[L - Co(l - pn)_3]^{3+}$.

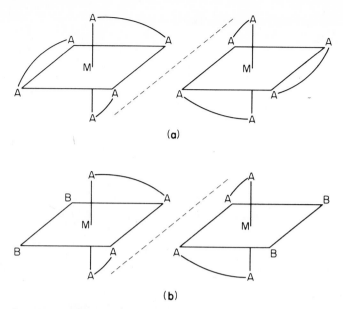

Fig. 7.8 Optical isomers for (a) $M(AA)_3$ and (b) $M(AA)_2B_2$.

7.6 LIGAND FIELD STABILIZATION ENERGY[3]

It is often necessary to determine which stereochemistry a metal ion will adopt in any particular situation. For example, in a discussion of ligand exchange reactions, it is important to know what stereochemistry is favored by the transition state. A parameter that is a guide to which stereochemistry forms is the ligand field stabilization energy (LFSE), which estimates the overall stabilization of the d orbitals brought about by the effect of the ligand field. An octahedral field with the d-electron configuration $(t_{2g})^p(e_g)^q$ has the stabilization

$$(-4p + 6q)Dq$$

Similar equations for other stereochemistries enable us to compare the relative LFSE for various shapes, given in Table 7.7.

By considering the difference between the LFSE for one stereochemistry and another, we can suggest which shape is the most likely to be found for a particular electron configuration. On the basis of such arguments, tetrahedral complexes are not expected to be common; in the past, it was considered that d^8 tetrahedral complexes would be unlikely. However, a number of compounds with tetrahedral stereochemistry are now known for Ni(II)(d^8). This points out that deductions from LFSE can be in error and

Table 7.7   Crystal Field Stabilization Energy in Units of Dq

	Octahedron		Tetrahedron	Square plane		Square pyramid		Pentagonal bipyramid	
	W*	S*	W	W	S	W	S	W	S
d^0	0	0	0	0	0	0	0	0	0
d^1	4	4	2.67	5.14	5.14	4.57	4.57	5.28	5.28
d^2	8	8	5.34	10.28	10.28	9.14	9.14	10.56	10.56
d^3	12	12	3.56	14.56	14.56	10.00	10.00	7.74	7.74
d^4	6	16	1.78	12.28	19.70	9.14	14.57	4.93	13.02
d^5	0	20	0	0	24.84	0	19.14	0	18.30
d^6	4	24	2.67	5.14	29.12	4.57	20.00	5.28	15.48
d^7	8	18	5.34	10.28	26.84	9.14	19.14	10.56	12.66
d^8	12	12	3.56	14.56	24.56	10.00	10.00	7.74	7.74
d^9	6	6	1.78	12.28	12.28	9.14	9.14	4.93	4.93
d^{10}	0	0	0	0	0	0	0	0	0

* W = Weak field; S = Strong field.

for this reason must be treated cautiously. It is often forgotten that the LFSE is at the most only 10% of the total energy of the system. In the case of nickel, complexes are often prepared in aqueous solution where the stable $Ni(H_2O)_6^{2+}$ ion dominates reducing the chance of preparing tetrahedral complexes. In the right medium and by using weak ligands, it is possible to prepare tetrahedral Ni(II) complexes.

For substitution in octahedral complexes, it is suggested that the transition state may be either five coordinate (probably square pyramid) or seven coordinate (pentagonal bipyramid). If this is so, then the data in Table 7.7 indicate the greatest loss of LFSE occurs for the d^3, d^6 spin paired, and d^8 configurations. This suggests that metal ions with these configurations are the least reactive, which agrees with observation.

BIBLIOGRAPHY

1. C. J. BALLHAUSEN, *Introduction to Ligand Field Theory*, McGraw-Hill Book Company, New York, 1962.

2. C. J. BALLHAUSEN and H. B. GRAY, *Molecular Orbital Theory*, W. A. Benjamin, Inc., New York, 1965.

3. F. BASOLO and R. G. PEARSON, *Mechanisms of Inorganic Reactions*, 2nd ed., John Wiley & Sons, Inc., New York, 1967.

4. R. S. NYHOLM, *Proc. Chem. Soc.*, 273, (1961).

5. B. E. DOUGLAS and D. H. McDANIEL, *Concepts and Models in Inorganic Chemistry*, Blaisdell Publishing Co., New York, 1965.

6. R. C. WILKINS, *Modern Coordination Chemistry*, ed. by J. Lewis and R. C. Wilkins, Interscience Publishers, New York, 1960.

FURTHER READING

F. A. COTTON and G. WILKINSON, *Advanced Inorganic Chemistry*, 3rd ed., Interscience Publishers, New York, 1972.

L. E. ORGEL, *Transition Metal Chemistry*, 2nd ed., Methuen & Co., Ltd., London, 1966.

T. M. DUNN, D. S. MCCLURE, and R. G. PEARSON, *Crystal Field Theory*, Harper, Row and Weatherhill, Tokyo, 1965.

B. N. FIGGIS, *Introduction to Ligand Fields*, Interscience Publishers, New York, 1966.

D. SUTTON, *Electronic Spectra of Transition Metal Complexes*, McGraw-Hill Book Company, New York, 1968.

8

Coordination Number and Stereochemistry

In this chapter, we consider the factors that influence the coordination number of atoms. The coordination number is distinct from stereochemistry and is the number of atoms bonded to a central atom.

8.1 IONIC COMPOUNDS

Ionic compounds, in the solid state, consist of an infinite array of ions in a crystal lattice. The coordination number of any ion in a lattice is determined principally by the relative size of the cation and anion. If the ions are considered as perfect spheres, then the problem of the coordination number becomes one of packing the spheres together. The most stable arrangement occurs when the surrounding ions barely touch the central one, of opposite charge, and barely touch one another. The coordination number therefore becomes a function of the radius ratio r_{A^+}/r_{X^-} for ions A^+ and X^-. Table 8.1 lists radius ratio limits for various coordination numbers. The radius ratio rule is equally applicable to monoatomic ions, as in NaCl, and to polyatomic ions, as in $CaCO_3$. This rule is limited in usefulness because accurate values for r_{A^+} and r_{X^-} are often not available, particularly for polyatomic ions.

Table 8.1   Radius Ratio

r_{A^+}/r_{X^-}	Coordination number	Shape
0.155–0.225	3	Triangular
0.225–0.414	4	Tetrahedral
0.414–0.732	6	Octahedral
> 0.732	8	Cubic

The arrangement of ions in a crystal lattice, the subject of crystal chemistry, will not be discussed further in this text. Our concern is with molecules and ions that are predominantly covalent in character.

8.2   COORDINATION NUMBER

It is becoming increasingly common to call any atom, or polyatomic species, bonded to a central atom a *ligand*. For example, the fluorine atoms in SF_6 are termed ligands just as the ammonia molecules are in $[Co(NH_3)_6]^{3+}$. This convention will be used throughout the discussion.

Ligand and Central Atom Size

The maximum stable coordination number of an atom is influenced by ligand size. The ion FeF_6^{3-} is stable, whereas $FeCl_6^{3-}$ only exists under the stabilizing influence of a very large cation. The more common chloro- complex is the tetrachloro- species $FeCl_4^-$.

Coordination numbers greater than six are more common for the larger central atoms, i.e., elements of the third and fourth long periods. For example, eight-coordinate cyanide complexes are known only for the heavy transition metals molybdenum, tungsten, and rhenium. For the same reason, there is a distinct trend in increasing coordination number as one moves down a group; the preferred chloro- complexes of aluminum and gallium are MCl_4^-, whereas those of indium and thalium are six-coordinate MCl_6^{3-}.

Oxidation State of the Central Atom

The oxidation state of the central atom has a decisive influence on its coordination number, especially for the *p* block elements. If the central atom is in an oxidation state of four, it will normally bond four monovalent ligands or two divalent ligands to give a neutral compound. However, the coordination number can be increased by polymerization or the formation of charged species. For example, the coordination number of aluminum is four in Al_2Cl_6 and in the anion $AlCl_4^-$, whereas in crystalline $AlCl_3$ it is six.

As an example of the influence of oxidation state, we shall consider sulfur (VI). It forms a six-coordinate compound SF_6 and a three-coordinate species SO_3, but in the latter case the coordination number can be increased to four, as in the sulphate anion. The additional charge placed on the sulfur atom in SO_4^{2-} can be accommodated for two reasons: the actual charge transfer from oxygen to sulfur is only partial; the charge can be dispersed over the anion and is not concentrated on the sulfur atom.

Electroneutrality

The electroneutrality principle states that the transfer of charge between bonded atoms is such as to make the residual charge on each of the atoms close to zero. Therefore, the formal charge on the central atom and the amount of electron transfer from the ligands can influence the coordination number. This is demonstrated for the complexes formed by divalent zinc—viz., $Zn(H_2O)_6^{2+}$ and $ZnCl_4^{2-}$. For the hydrate, the charge transfer from each water molecule is approximately one-third electron; thus, Zn^{2+} can coordinate six water molecules. In the chloro-anion $ZnCl_4^{2-}$, the charge transfer is approximately one-half electron per chlorine; hence, the coordination number is restricted to four. The coordination number in metal ammine complexes can be, in a number of cases, related to such an electroneutrality principle. Monovalent cations generally coordinate two ammonia molecules, divalent cations coordinate four, and trivalent cations coordinate six, but there are exceptions to this rule.

The isoelectronic ions Ag^+, Cd^{2+}, and In^{3+} preferentially form the chloro- complex anions $AgCl_2^-$, $CdCl_4^{2-}$, and $InCl_6^{3-}$. This change in coordination number can be directly related to the oxidation state of the metal and the electroneutrality principle.

Electronegativity of Ligand

Highly electronegative ligands favor high coordination numbers. This is a factor contributing to the formation of SF_6. The electronegative fluorine atoms increase the positive character of the S atom, which is believed to cause contraction of the d orbitals. The contracted d orbitals are now more efficient in overlap with the F $2p$ orbital, and therefore the energy expended in producing the sulfur valence state $3s^1 3p^3 3d^2$ will be recovered by the formation of six strong bonds. Chlorine, with a lower electronegativity, cannot do this, and the highest coordination number achieved with sulfur is four in SCl_4.

Orbital Energies

There is a marked discontinuity in coordination number when we progress from the first short period to the later periods, because the elements of the first short period have only $2s$ and $2p$ orbitals of favorable energy for bonding. The heavier elements have accessible d orbitals and can expand their valence shell.

The influence of the availability of additional bonding orbitals can be demonstrated for the elements Cu, Ag, and Au.[1] The valence shell s-p separation for the free metal atoms is 3.79, 3.66, and 4.63 eV, respectively (Fig. 8.1). This means that the energy required to involve the s and p orbitals

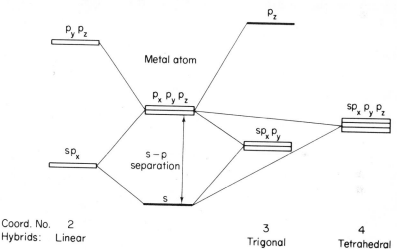

Coord. No.	2	3	4
Hybrids:	Linear	Trigonal	Tetrahedral

Fig. 8.1   Linear, trigonal, and tetrahedral hybridization schemes relative to the s-p energy separation.

of gold in bonding is greater than that for silver or copper. It is common, therefore, to find that most Au(I) compounds are two coordinate and linear, whereas four-coordinate compounds more frequently form for Cu(I) and Ag(I). However, a careful choice of ligand may force the Au(I) to become four coordinate, as in $[Au(diarsine)_2]I$.

Bond Type

The ability of the central atom and the ligands to form multiple bonds can affect the coordination number. For example, when the positive charge on a transition metal is low (i.e., low oxidation state), its capacity for accepting electrons from the ligands is limited. However, if the excess negative charge can be dispersed through back bonding (dative $M \rightarrow L$ π bond), then the number of coordinating ligands may be increased. The stability of the hexacyanides of Fe(II) and Fe(III) may result from extensive π bonding.

Nine-Orbital Rule

If d orbitals can be included in the bonding orbitals of a central atom, then the maximum number of bonding orbitals is nine $s^1p^3d^5$. This nine-orbital rule is another way of stating the inert gas configuration or effective atomic number rule, which is, that on forming compounds elements try to achieve the electron configuration of the nearest inert gas. For

Table 8.2 Nine-Orbital Rule and Maximum Coordination Number

Configuration	Number of unpaired electrons	Maximum coordination number (based on nine-orbital rule)	Examples
d^0		9	ReH_9^{2-}
d^1	1	8	$Mo(CN)_8^{3-}$
d^2	2	7	
	0	8	$Re(diarsine)_2Cl_4^+$
			$Mo(CN)_8^{4-}$
d^3	3	6	$CrCl_6^{3-}$
	1	7	$Mo(CN)_7^{4-}$
d^4	4	5	$MnCl_2(diarsine)H_2O^+$
	2	6	$Re(diarsine)_2Cl_2^+$
	0	7	Re in $Re_3Cl_{12}^{3-}$
d^5	5	4	$MnCl_4^{2-}$
	3	5	$[Mn(sal-Me)_2]_2$?
	1	6	$Mn(CN)_6^{4-}$
d^6	4	4	$FeCl_4^{2-}$
	2	5	$Co(8\text{-aminoquinoline})_2NO^{2+}$
	0	6	$PtCl_6^{2-}$
d^7	3	4	$CoCl_4^{2-}$
	1	5	$NiBr_3(Et_3P)_2$
d^8	2	4	$NiCl_4^{2-}$
	0	5	$Fe(CO)_5$
d^9	1	4	$Cu(NH_3)_4^{2+}$
d^{10}	0	4	$Zn(CN)_4^{2-}$

example, oxygen in the water molecule achieves the neon configuration, whereas the hydrogens achieve the helium electron configuration.

For the case of the d^2 configuration of a transition metal when the electrons remain unpaired, the maximum coordination number possible is seven, and if the two electrons are paired, the maximum increases to eight. Table 8.2 lists the maximum coordination numbers on the basis of the nine-orbital rule. Some of these coordination numbers have not yet been realized, as the last column of the table indicates. However, in general, we can say that to obtain high coordination numbers we use metal ions with a small number of d electrons, whereas for low coordination numbers the metal ion is more likely to have a lot of d electrons. Coordination numbers greater than those given in Table 8.2 are readily achieved for configurations d^4 and greater. This probably occurs through placing some d electrons in antibonding orbitals, which leads to the central ion obtaining more electrons than predicted by the nine-orbital rule.

The erratic coordination numbers found for simple metal carbonyls are readily explained by the nine-orbital rule, or 18-electron rule, as it is sometimes called (Table 8.3). The hexacarbonyl of vanadium $V(CO)_6$ is an

Table 8.3   Metal Carbonyls

	$Cr(CO)_6$	$Mn_2(CO)_{10}$	$Fe(CO)_5$	$Co_2(CO)_8$	$Ni(CO)_4$	(Kr)
Metal electrons	24	25	26	27	28	(36)
Electrons required	12	11	10	9	8	
Method	6(CO)	5(CO)	5(CO)	4(CO)	4(CO)	
		+(Mn—Mn) bond		+(Co—Co) bond		

exception; on the basis of the rule, it should be formulated as $V_2(CO)_{12}$, with a V—V bond. However, this would mean seven coordination around the vanadium, which is probably sterically difficult.

Each of the seven factors discussed has been considered as independent effects, but in any one compound it is not always easy to disentangle them or even decide on the dominant one.

BIBLIOGRAPHY

1. R. S. NYHOLM, *Proc. Chem. Soc.*, 273 (1961).

PART II

9

First Short Period

Li   Be   B   C   N   O   F   Ne

The elements of the first short period will be considered separately from the heavier elements of the same periodic groups, because (a) the maximum number of valency electrons they can attain is eight, restricting the coordination number to four, and (b) the elements C, N, and O can take part in extensive multiple bonding. Some important properties of the elements are given in Table 9.1. The elements can accommodate a total of 10 electrons $(1s)^2(2s)^2(2p)^6$, but to a first approximation the $(1s)^2$ electrons can be neglected in properties relating to chemical valency. The four remaining electron pairs are expected to be directed tetrahedrally (Chapter 4). This has been confirmed experimentally from the structure of crystalline neon, which has a cubic close packed structure with each neon atom surrounded by four others in a tetrahedral array.

The chemistry of the first short period elements is dominated by the attainment of the filled valency shell $(2s)^2(2p)^6$.

Table 9.1 Some Properties of the Elements of the First Short Period

Element	Li	Be	B	C	N	O	F	Ne
Valence electron configuration	$(2s)^1$	$(2s)^2$	$(2s)^2(2p)^1$	$(2s)^2(2p)^2$	$(2s)^2(2p)^3$	$(2s)^2(2p)^4$	$(2s)^2(2p)^5$	$(2s)^2(2p)^6$
Ionic radius* (pm)	Li^+ 60	Be^{2+} 31	B^{3+} 20	C^{4+} 15 C^{4-} 260	N^{5+} 11 N^{3-} 171	O^{6+} 9 O^{2-} 140	F^{7+} 7 F^- 136	
Covalent radius (pm)	133	89	88	77	70	66	64	112
Atomic radius (pm)			80	77	74	74	72	160
First ionization potential (eV)	5.39	9.32	8.296	11.264	14.54	13.614	17.418	21.559
Electron affinity (eV)	0.54	~0	0.3	1.13	~-0.2	1.48	3.62	0
Electronegativity*	0.97	1.47	2.01	2.50	3.07	3.50	4.0	
Principal valency state	1	2	3	4	3	2	1	
Typical compounds	LiCl LiO_2	$BeCl_2$ $BeSO_4$	BCl_3 B_2H_6 $B(OH)_3$	C_3H_8 CCl_4 C_2H_5OH	NH_3 HNO_3 N_2H_4	H_2O SiO_2 KOH	PF_3 KF ClF_3	

* After Pauling.

9.1   CHEMICAL AND PHYSICAL PROPERTIES

It is necessary to consider briefly some of the more important physical and chemical properties of the elements of the first short period, since these properties provide clues to the coordination numbers and stereochemistry of the elements.

The Physical State of the Elements and Their Compounds

It is apparent from Table 9.2 that there is a marked change (around boron and carbon) in the physical properties of the elements and their compounds; this can be understood in terms of the electronic configuration of the elements. In monomeric covalent compounds of Li, Be, and B, the maximum of eight valence electrons has not been achieved. For Be and B, polymerization is common, producing compounds (e.g., oxides) with extended bonding throughout the solid. On the other hand, C, N, O, and F can, in their normal valency state, achieve the inert gas configuration of neon by forming monomeric compounds, and polymerization is not necessary. The same applies in the free elemental state; for example, in the case of nitrogen and oxygen the formation of multiple bonds to give $N\equiv N$ and $O=O$ allows the atoms to achieve the neon electron structure. An alternative single bond arrangement for elemental nitrogen and oxygen would lead to polymeric structures containing lone pairs in close proximity. Owing to the small size of the N and O atoms, the resulting repulsions could well destabilize any extended single bond structure. Lone pair repulsions are given as the reason for the weakness of the $N-N$ (161 kJ mol^{-1}), $O-O$ (146 kJ mol^{-1}), and $F-F$ (158 kJ mol^{-1}) single bonds. Oxygen achieves some measure of polymerization in the allotrope ozone, O_3, but even here some multiple bonding occurs.

Multiple Bonding

The elements readily form multiple bonds either by normal electron sharing or by dative electron-pair donation. The multiple bonds in the entities $>C=C<$, $-C\equiv C-$, $>C=O$, $N\equiv N$, and $O=O$ are of the first type; the second type of π bond occurs for the elements Be and B in compounds such as monomeric BeX_2 and BX_3. In BF_3, for example, the boron atom has an unfilled p orbital, perpendicular to the plane of the molecule, which can accept a fluorine lone pair in a dative $F \rightarrow B$ π bond.

Table 9.2 Properties of Some Compounds of the Elements of the First Short Period

Elements	Li	Be	B	C	N	O	F
Physical state	Solid	Solid	Solid	Solid	Gas	Gas	Gas
Structure	Body c.c. and h.c.p.	h.c.p. 12 nearest neighbors	Nearly regular icosahedron of B_{12} group	Tetrahedral (diamond) layer, hexagonal rings of carbons (graphite)	$N{\equiv}N$ ←——— diatomic ———→	$O{=}O$	$F{-}F$
Melting point (°C)	179	1285	2200	3850	−201.5	−218.9	−223
Oxides	Li_2O	BeO	B_2O_3	CO_2	N oxides	O_3	OF_2
Structure	Antifluorite	Wurtzite	Probably contain BO_4 tetrahedral units	$O{=}C{=}O$		$O{=}O{\cdots}O$	$F{-}O{-}F$
Melting point (°C)	1700	2570	~400	Gas	Gases	Gas	Gas

Fluorides	LiF	BeF$_2$	BF$_3$	CF$_4$	NF$_3$	OF$_2$	
Structural type	NaCl structure Ionic solid	Polymeric BeF$_4$ tetrahedral groups Solid	Monomeric Trigonal planar Gas	Monomeric Tetrahedral Gas	Monomeric Pyramidal Gas	Monomeric Angular Gas	
Melting or boiling point (°C)	842 (mp)	803 (bp)	−101 (bp)	−128 (bp)	−129 (bp)	−145 (bp)	

Chlorides	LiCl	BeCl$_2$	BCl$_3$	CCl$_4$	NCl$_3$	OCl$_2$	ClF
Structural type	NaCl structure Ionic solid	Polymeric BeCl$_4$ tetrahedral units Solid	Monomeric Trigonal planar Liquid	Monomeric Tetrahedral Liquid	Monomeric Pyramidal Liquid	Monomeric Angular Liquid	Monomeric? Linear Gas
Melting or boiling point (°C)	614 (mp)	405 (mp)	12.5 (bp)	76 (bp)	~71 (bp)	3.8 (bp)	−101 (bp)

The formation of such a π bond for BF_3 can occur only if the molecule is planar.

Oxidation States

The oxidation state of the elements is important in determining the *minimum* covalency around the central atom. The principal oxidation state is N or $8 - N$ (whichever is the smaller), where N is the group number to which an element belongs in the periodic table (i.e., the number of electrons in the valence shell). Boron is therefore trivalent and compounds BX_3 and B_2X_3 would be expected for monovalent and divalent ligands, respectively. Since beryllium and boron are 4 and 2 electrons short of the completed octet in compounds BeX_2 and BX_3, respectively, these elements in such compounds can act as electron acceptors. The elements N, O, and F, on the other hand, in forming compounds NX_3, OX_2, and FX (X is monovalent), achieve the complete octet but have 1, 2, and 3 lone pairs, respectively, and therefore can act as electron donors. The donor and acceptor properties provide a method by which the central atom can increase its coordination number above that determined by the oxidation state.

9.2   COORDINATION NUMBER AND PHYSICAL STATE

The range of coordination numbers exhibited by the elements is given in Table 9.3. The coordination number of Be in $BeCl_2$ is 2, 3, or 4, depending

Table 9.3   Coordination Numbers

Li	Be	B	C	N	O	F	Ne
			1	1	1	1	
	2		2	2	2	2	
	3	3	3	3	3		
4	4	4	4	4	4	4	(4)
6			(5)	(6)			

on the physical state of the compound. The coordination number 4 occurs in solid $BeCl_2$, whereas the lower coordination numbers occur in the gaseous state. In general, where the coordination number of an element in a particular compound varies, the lower coordination occurs at the higher

Table 9.4 Coordination Number and Physical State

LiF	$BeCl_2$	$BF_3 \cdot SMe_2$	H_2O
R.T.	R.T.	$< -20°$	$0°$
Ionic solid	Solid	Solid	Solid
CN = 6 (Li)	CN = 4 (Be)	CN = 4 (B)	CN = 4 (O)
Vapor	$564°$	$> 50°$	$0–100°$
$(LiF)_3$	23% Be_2Cl_4	~95%	Liquid
		BF_3, SMe_2	
$(LiF)_2$	CN = 3 (Be)	CN = 3 (B)	CN = 4, 3, 2 (O)
LiF	$1000°$		$> 100°$
	100% $BeCl_2$		Gas
CN = 1 (Li)	CN = 2 (Be)		CN = 2 (O)
Ionic lattice	Polymerization	Dative bonding	H-bonding

temperature (Table 9.4) as increased thermal energy causes bond cleavage of certain of the bonds. Similarly, when an element can increase its coordination number by electron acceptance or donation, the ability to do so is often temperature dependent. It is therefore important to specify the physical state of a compound when discussing the coordination number and stereochemistry of an element. As a consequence of coordination number changes, both the bond angles and bond lengths may vary significantly with changes in temperature or state (Table 9.5).

Table 9.5 Beryllium Dichloride

Physical state	Coordination number of Be	ClBeCl angle	Be—Cl length (pm)
Solid	4	$98°$ (bridging Cl's)	202
Gas (1000°)	2	$180°$	177

9.3 STEREOCHEMISTRY OF THE FIRST SHORT PERIOD ELEMENTS

The remainder of the chapter will be concerned with a general discussion of the stereochemistry of the elements, followed by a consideration of each shape assumed by the elements in greater detail.

Table 9.6 Stereochemistry of the Elements of the First Short Period*

Coordination number	Be	B	C	N	O	F
	$(2s)^2$	$(2s)^2(2p)^1$	$(2s)^2(2p)^2$	$(2s)^2(2p)^3$	$(2s)^2(2p)^4$	$(2s)^2(2p)^5$
1			:C≡	:N≡	$\overset{\cdot\cdot}{\text{O}}$=	$\overset{\cdot\cdot}{\underset{\cdot\cdot}{\text{F}}}$—
				$\overset{\cdot\cdot}{\text{N}}$=		
					:$\overset{\cdot\cdot}{\underset{\cdot\cdot}{\text{O}}}$—	
				:$\overset{\cdot\cdot}{\underset{\cdot\cdot}{\text{N}}}$—		
2	=Be=		—C≡	=N=	=O=	—$\overset{\cdot\cdot}{\underset{\cdot\cdot}{\text{F}}}$—
			=C=			
3	==Be⟨	==B⟨	=C⟨	$\overset{\cdot\cdot}{\text{N}}$⟨	$\overset{\cdot\cdot}{\text{O}}$⟨	$\overset{\cdot\cdot}{\text{F}}$⟨
			—C⟨	==N⟨		
			==C⟨	=N⟨		
4	Be	B	C	N	O	F

* Lone pairs are represented by **:**

Shapes	—*X*— linear	*X* pyramidal	*X* tetrahedral
	X bent	—*X*⟨ trigonal planar	

The stereochemistries displayed by the elements are summarized in Tables 9.6 and 9.7, which show that the dominant factor is the attainment of the filled valence shell of four electron pairs and their subsequent arrangement in space. For example, the sum of the numbers in the second, third, and fourth columns across Table 9.7 is always four. Therefore, the basic stereochemical problem is simply the various possible arrangements of four

Table 9.7¹ Stereochemistry of the Elements of the First Short Period

Total no. σ pairs	Bonding σ pairs	π bonds	Lone pairs	Bond arrangement	Symmetry (of bonds)	Element	Examples
4	4	0	0	Tetrahedral	T_d	Be, B, C, N, O, (F)*	$BeCl_{2(s)}$, $BF_3 \cdot Et_2O$, CH_4, NH_4^+, $(Be_4\ acetate_6)O$, NH_4F
	3	0	1	Pyramidal	C_{3v}	N, O, (F)	NH_3, H_3O^+, $N_2H_6F_2$
	2	0	2	Angular / Linear	C_{2v}	N, O, F / (F)	NH_2^-, H_2O / PdF_3 (bridging F)
3	3	1	0	Trigonal planar	C_{3v}	Be, B, C, N	$BeCl_{2(g)}$, BF_3; C_6H_6, NO_3^-
	2	1	1	Angular	C_{2v}	N, O	NOF, O_3
2	2	2	0	Linear		Be, C, N, O	$BeCl_{2(g)}$; CO_2, HN_3, C_2H_2, CN^-
	1	2	1	Diatomic		C, N	$[Ru_2Cl_{10}O]^{4-}$, CO
	1	1	2	Diatomic		N, O	O_2

* (F) are for compounds in which fluorine is taking part in H bonds.

electron pairs, which may occur as σ, π, or lone pairs. The relationship between the various stereochemistries is depicted in Table 9.8. Five basic shapes exist: tetrahedral, trigonal plane, pyramidal, bent, and linear. The number of lone pairs increases from left to right, with a small decrease in bond angles owing to the steric influence of the lone pairs.

The expected limits of the bond angles corresponding to a particular bond arrangement, based on available bonding orbitals and the Pauli exclusion principle, are also listed in Table 9.8. More precise calculations,

Table 9.8   Stereochemistry of the Elements of the First Short Period*

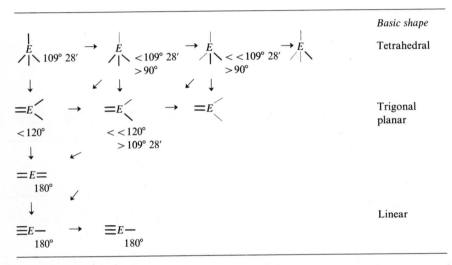

```
* —    Bonding pairs.
  —    Lone pairs.
  =    Multiple bonds.
  <    Less than.
  < <  A further degree smaller, not "much less than."
```

using the method of configuration of maximum probability (Table 4.12, Section 4.8), lie within the predicted ranges.

Another way of approaching the stereochemical relationships listed in Table 9.8 is to consider elements diagonally across the table (right to left), so that the coordination number remains unchanged. This approach, which indicates the extremes of shape that can occur for the same coordination number, is the basis of the method used by Walsh to describe stereochemistry (Section 5.3).

9.4 FOUR COORDINATION, TETRAHEDRAL STEREOCHEMISTRY

Tetrahedral stereochemistry occurs for all four coordinate compounds of the first short period elements. Table 9.9[2] lists a few of these compounds with this stereochemistry. Only when the four ligands are identical and non-

Table 9.9   Four-Coordinate Compounds

Element	Compound	Bond length (pm)		Bond angle	
Be	$(BeCl_2)_n$		202		98°
	$Be(acac)_2$				109°
	$Be_4O'(O_2CCH_3)_6$	BeO	167	OBeO	102° 7'
		BeO'	162	OBeO'	115.2°
	$[Be(CH_3)_2]_n$		193		114°
B	$(CH_3)_3N \cdot BF_3$	BN	158.5	FBN	112°
		BF	139	FBF	107°
	$BH_3 \cdot CO$	BC	154		
		BH	120		
	$BaBOF_3$	BF	143		
		BO	143.2		
C	(see Table 9.10)				
N	$(CH_3)_3NBF_3$	CN	150	CNC	114°
		BN	158.5		
	CH_3NO	NO	136		
		CN	149	CNC	114°
O	PdO			PdOPd	98°
	Oxides M_2O cuprite ⎫				
	MO zinc blende ⎪			Tetrahedral	
	M_2O_3 corundum ⎬			OM_4 groups	
	MO_2 fluorite ⎭				
	$LiClO_4 3H_2O'$	Li—O'	215	LiOLi	102°
				HOH	130°

bridging is the stereochemistry strictly regular. There are various reasons for distortions from the regular stereochemistry. For example, the elongation in $(BeCl_2)_n$ has been attributed to repulsion between adjacent beryllium atoms. In boron compounds of the type BX_3L (e.g., $BF_3 \cdot CH_3CN$), the FBL bond angle is greater than 109° 28'. Repulsion between the bonding electrons in the B—F bonds is probably less than between B—F and B—L bonds, owing to the greater electronegativity of fluorine. The closely related carbon compounds listed in Table 9.10 have bond angles close to 109° 28'. However, the small observed trends may be accounted for in terms of the electronegativity and size of the halide ligands. The $C-X$ bonds are longer than the $C-H$ bonds, and since the electronegativity of hydrogen is less than that of F, Cl, and Br, the maximum electron density in the $C-X$ bonds is further from carbon than it is in the $C-H$ bonds. We find, therefore, that the HCH bond angles in CH_3X and CH_2X_2 are greater than 109° 28', because there is greater repulsion between the $C-H$ bonds than between the $C-X$ bonds.

Table 9.10   Bond Angles in Tetrahedral $CH_{4-x}X_x$ Compounds

Angle		CH_4	CH_3X	CH_2X_2	CHX_3	CX_4
	X					
HCH	F	109° 28'	110° 0'	112°		
	Cl	109° 28'	110° 20'	112°		
	Br	109° 28'	110° 48'			
	I	109° 28'	110° 58'			
XCX	F			108°	108° 48'	109° 28'
	Cl			111° 47'	110° 24'	109° 28'
	Br				110° 48'	109° 28'
	I					109° 28'

The same reasoning would lead us to expect that the XCX bond angles for the compounds CHX_3 and CH_2X_2 would be less than 109° 28', but this is only so for CHF_3 and CH_2F_2. The Cl, Br, and I atoms are larger than F and H, and the electronegativity effect seems to be offset by steric interactions.

The tetrahedral environment of nitrogen in the compounds $(CH_3)_3N-BF_3$ and $(CH_3)_3N-O$ is distorted, with the CNC angle 6° greater than in pyramidal $(CH_3)_3N$. The replacement of the lone pair in $(CH_3)_3N$ by the N—B or N—O bonding pairs removes the steric influence of the lone pair and allows the methyl groups to move further away from each other.

9.5   THREE COORDINATION

Three coordinate compounds of the first short period elements have either a pyramidal or trigonal planar structure. The latter arrangement occurs for Be, B, C, and N and the former for C, N, O, and F. The difference lies in the use made of the fourth valence orbital; if it is filled or empty and involved in π bonding, or is empty and not used in π bonding, the stereochemistry is trigonal planar; otherwise, the compounds are pyramidal.

Pyramidal Stereochemistry

The pyramidal stereochemistry is most common for nitrogen (Table 9.11). The bond angle XNX in the majority of compounds NX_3 and NX_2Y is close to 108°, the value predicted by the method of configuration of maximum probability. However, this stereochemistry can be upset by

Table 9.11   Three-Coordinate Pyramidal Compounds

Compound	Bond length (pm)		Bond angle	
NH_3		101.5		106.6°
NF_3		137.1		102.1°
$N(CF_3)_3$		143		114°
NH_2Cl	NH	101.4	HNH	106.8°
	NCl	177	HNCl	102°
$N(CH_3)_3$		147		108°
$N(CH_3)_2H$	CN	146	CNC	108°
NH_2OH	NO	146	HNO	105–107°
	NH	101–103	HNH	107°
$N(CH_3)_2Cl$	CN	147	CNCl	107°
	NCl	177	CNC	108°
$(C_2H_5)_2NH$	CN	147	CNC	112°
N_2H_4	NH	102	NNH	112°
	NN	145		
$(CH_3)_2N_2H_2$	CN	147	CNC	110°
	NN	145	CNN	110°
	NH	104		
N_2F_4	NN	147	FNF	108°
	NF	137	NNF	104°
H_3O^+ (in $HClH_2O$)		96		117°
$(CH_3)_2OBF_3$	OB	150		109.5°
	CO	145		

other factors. For example, the bond angle in NF_3 is considerably less than 108°, which has been attributed to the high electronegativity of fluorine. The polar nature of the N—F bond is indicated by the low dipole moment of NF_3 (0.23 D), where the polarity almost balances the effect of the lone pair. A bond angle of 114° for $N(CF_3)_3$, on the other hand, may be due to fluorine nonbonded electron repulsions or to a hyperconjugative effect.

The nitrogen atoms are pyramidal in hydrazine and its derivatives with a XNN bond angle around 110 to 112°. The high dipole moment of hydrazine suggests that it exists in the gauche configuration, and that rotation about the N—N bond is prevented by the lone pair on each nitrogen. The lone pair repulsions are probably responsible for the long N—N bond (145 pm). The bond is shortened when the lone pairs are involved in bonding, as in $N_2H_5^+$ and $N_2H_6^{2+}$.

Trigonal Planar Stereochemistry

Trigonal planar compounds of carbon and nitrogen contain three σ and one π bond, and, for beryllium and boron, three σ and possibly one π bond. The π bond in both cases can be either localized in one direction or delocalized in two or three directions. Information on bond lengths (Table 9.12) or a plot of bond length against bond order (Fig. 9.1) indicates

Table 9.12 Bond Length and Bond Order

Bond	Compound	Bond length (pm)	Bond order
C—C	Ethylene	134	2
	Benzene	139	1.5
	Graphite	142	1.33
C—C Single bond		154	1
C—O	CH_3CHO	122	2
	$C_2O_4^{2-}$	125	1.5
	CO_3^{2-}	129	1.33
C—O Single bond		143	1

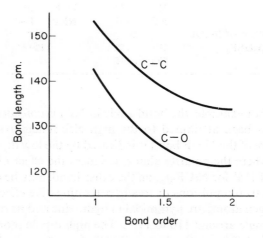

Fig. 9.1 Bond length plotted against bond order for C—C and C—O bonds.

the degree of delocalization. Bond lengths can therefore be used as a guide to bond multiplicity. However, it is necessary to be cautious in such interpretations as bond lengths can also depend on the percentage of s and p character of a bond. For example, the $C{=}O$ bond lengths in compounds $X_2C{=}O$ change with X, and this can be correlated with the change in s character in the $C{-}O$ σ bond (Table 9.13). Changes in the $C{-}C$ bond

Table 9.13   Bond Length and Percent *s* Character

Compound	C=O length (pm)	Calculated % s character
$\begin{array}{c}H\\ \diagdown\\ \quad C{=}O\\ \diagup\\ H\end{array}$	123.0	26
$\begin{array}{c}Cl\\ \diagdown\\ \quad C{=}O\\ \diagup\\ Cl\end{array}$	116.6	46
$\begin{array}{c}F\\ \diagdown\\ \quad C{=}O\\ \diagup\\ F\end{array}$	117	44

lengths can be treated in a similar way (Table 9.14). From Table 9.14, we see that the bond shortening owing to the π delocalization in benzene, for example, is 6 pm and not 14 pm; that is, the length of the $C{-}C$ bond in benzene should be compared with an $sp^2{-}sp^2$ single bond and not an $sp^3{-}sp^3$ single bond. Even this is not a valid comparison, inasmuch as the $C{-}C$ bond length in $\gtrsim C{-}C\lesssim$ is influenced by conjugation. In adducts of BF_3, it is very likely that the observed increase in the $B{-}F$ bond length from BF_3 to BF_3L can be attributed to both a loss of the $B{-}F$ π bonding and a decrease in the s character of the $B{-}F$ bonds, but not to just one of these effects.[5]

Only a few trigonal planar compounds of beryllium are known; the dimeric compounds Be_2X_4, for example $[X = Cl, Br, I, CH_3, and (CH_3)_2CH]$, are postulated from infrared evidence. The phthalocyanine complex of beryllium also has a distorted trigonal planar stereochemistry; it is thought that only three of the four nitrogen atoms are bonded to the metal. The chemical observation that this complex readily forms a dihydrate, which is believed to contain two $Be{-}N$ and two $Be{-}OH_2$ bonds in a tetrahedral environment, supports this view.

Table 9.14[3,4]   Bond Lengths and Orbital Hybridization

C—C Hybridization	% s Character in σ bond	Bond arrangement	C—C Bond length (pm)
$sp^3 - sp^3$	25	$\underset{/}{\overset{\backslash}{-}}C-C\underset{\backslash}{\overset{/}{-}}$	154
$sp^3 - sp^2$	29	$\underset{/}{\overset{\backslash}{-}}C-C\underset{\backslash}{\overset{/\!/}{}}$	150
$sp^3 - sp$	33	$\underset{/}{\overset{\backslash}{-}}C-C\equiv$	146
$sp^2 - sp^2$	33	$\underset{/}{\overset{\backslash\!\backslash}{}}C-C\underset{\backslash}{\overset{/\!/}{}}$	146
$sp^2 - sp$	40	$\underset{/}{\overset{\backslash\!\backslash}{}}C-C\equiv$	142
$sp - sp$	50	$\equiv C-C\equiv$	138
$sp^2 - sp^2 + \pi$	33	$\underset{/}{\overset{\backslash}{}}C=C\underset{\backslash}{\overset{/}{}}$	134
$sp^2 - sp + \pi$	40	$\underset{/}{\overset{\backslash}{}}C=C=$	131
$sp - sp + \pi$	50	$=C=C=$	128
$sp - sp + 2\pi$	50	$-C\equiv C-$	120
$sp^2 - sp^2 + \frac{1}{2}\pi$	33	Benzene	140
$sp^2 - sp^2 + \frac{1}{3}\pi$	33	Graphite	142

The predominant stereochemistry of boron is trigonal planar (Table 9.15), and in such compounds boron can participate in π bonding as an electron acceptor. However, the π bond is not necessary for the planar shape, as it is in the case of three-coordinate planar nitrogen compounds.

A number of trigonal planar compounds of nitrogen are known, especially when the nitrogen is bonded to at least two oxygen atoms (Tables 9.15 and 9.16). We shall consider the bonding in one such species, the nitrate ion,[6,7] in more detail. If the z axis is taken as perpendicular to the molecular plane, the σ-bonding orbitals of the nitrogen are $2s$, $2p_x$, and $2p_y$ (Fig. 9.2). These will overlap with the $2p_x$ orbital of each oxygen (assuming, in this treatment, that the oxygen $2s$ electrons are nonbonding). The nitrogen atomic orbitals considered in the point group symmetry D_{3h} are $1a_1'$ ($2s$ type) and $1e'$ ($2p_x$, $2p_y$ type). The unused $2p_z$ orbital (a_2'') on the nitrogen and one

Table 9.15 Three-Coordinate Trigonal Planar Compounds*

	Compound		Bond length (pm)	Bond angle	
B	BF_3 ⎫		129.5	120°	
	BCl_3 ⎬ vapor		176	120°	
	BBr_3 ⎭		187	120°	
	$B(OH)_3$ cryst.		136	120°	
	$(BO_3)^{3-}$				
	in $Co_3(BO_3)_2$		135	121°	
	and $Mg_3(BO_3)_2$		143	119°	
	B_2Cl_4	BB	177	120°	
		BCl	173		
	B_2F_4	BB	167	FBF	115°
		BF	132	BBF	122.5°
	B_2Cl_4 ethylene	BC	158	ClBCl	116.6°
		BCl_{av}	175	CBCl	119.8°
					123.6°
	CH_3BF_2	CB	160	FBF	118°
		BF	130		
	$(C_6H_5)BCl_2$	CB	152	ClBCl	118°
		BCl	172		
C	(see Table 5.6 for examples)				
N	HNO_3	NO	122	ONO	130°
		NO—H	141		
	NO_3^-	NO	121.8	120°	
	$N(SiH_3)_3$	SiN	173.8	119.6°	
	$C(NO_2)_4$	CN	147	ONO	127°
		NO	122		

* See Table 9.16 for further examples.

Table 9.16 Three-Coordinate Trigonal Planar Compounds O_2N—X

Compound	ONO bond angle	N—X bond length (pm)	Comment
O_2N—O^-	120°	121.8	
O_2N—Cl	130° 35′	184	> N—Cl Single bond
O_2N—F	129.5°	135	~ N—F Single bond
O_2N—OF	125°	139	> N—O in NO_3^-
O_2N—OH	130°	141	> N—O in NO_3^-
O_2N—NO_2	126°	164	> N—N Single bond
O_2N—C_6H_5	124.5°	148	~ N—C Single bond
O_2N—CH_3	127°	147	~ N—C Single bond
O_2N—OCH_3	125°	136	> N—O in NO_3^-
O_2N—CCl_3	135°	159	> N—C Single bond
O_2N—NH_2	129°	140	~ N—N Single bond
O_2N—$N(CH_3)_2$	120°	130	< N—N Single bond
			> N=N Double bond

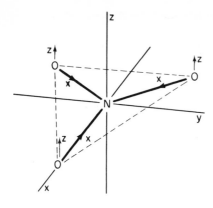

Fig. 9.2 Coordinate system for the NO_3^- ion.

on each oxygen have π symmetry and combine to give four π-molecular orbitals, one bonding $(1a_2'')$, two approximately nonbonding $(2a_2'', 3a_2'')$, and one antibonding $(4a_2'')$. The oxygen $2p_y$ orbitals also have π symmetry, but there are no corresponding orbitals available on the nitrogen; therefore, they remain nonbonding. The 24 valence electrons of the nitrate ion are accommodated in the bonding and nonbonding orbitals as follows (see Fig. 9.3):

$(2s2s2s)^6(2p_y2p_y2p_y)^6$	$(1a_1')^2(1e')^4$	$(1a_2'')^2$	$(2a_2''3a_2'')^4$
Nonbonding on the oxygens	σ bonding	π bonding	π non-bonding

Therefore, the bonding consists in three σ bonds, with one π bond delocalized over the ion. The influence of the π bond on the stereochemistry can be seen from the energy level diagram (Fig. 9.3). If the number of valence electrons drops below 12, there will be no π bond and a pyramidal shape results, as in NH_3; if there are more than 24 valence electrons, as in NF_3, the $4a_2''$ antibonding orbital will fill, the π bond is effectively removed, and the pyramidal shape will reoccur.

Compounds of the type NO_2—X have ONO bond angles in the range of from 125 to 135° (Table 9.16). A number of the ligands X are capable of taking part in π bonding. However, the N—X bond length in all the cases studied is close to the expected single bond length and, in a number of cases, even longer. In these compounds, the $1a_2''$ π-molecular orbital (Fig. 9.3) is composed of the $2p_z$ orbitals of the two oxygen atoms and the central nitrogen —that is, the π bond is now delocalized over two N—O bonds, not three as in the nitrate ion. This leads to additional electron density in the N—O bonds (of NO_2—X), and the corresponding increase in electron repulsion between these bonds increases the angle between them beyond 120°. A similar increase in the OCO angle is observed for carboxylate ions of the type RCO_2^-.

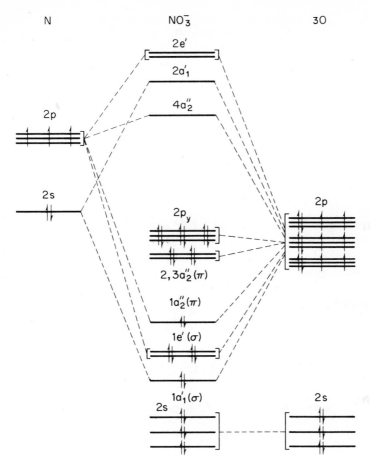

Fig. 9.3 Molecular orbital energy diagram for the NO_3^- ion.

9.6 TWO COORDINATION

The arrangement of electron pairs and the shapes of two-coordinate compounds of the first short period are given in Table 9.7, but not all species fall clearly into one of the three categories. Table 9.17 lists a range of such compounds and ions. Differences in the extent of π overlap, s-orbital character, and bond-bond repulsions probably all contribute to the final equilibrium shape. In fact, there seems little advantage in trying to distinguish between the three limiting shapes except in obvious cases. We can generalize as follows: (a) when bond angles are less than 109° 28′, the two bonds are probably single; (b) linear molecules occur when two π bonds are formed;

Table 9.17   Two-Coordinate Angular
Compounds

Compound	Bond angle	Bond length (pm)
Cl_2O	110.8°	170
$(CF_3)_2O$	125°	(136)
P_4O_6	127.5°	165
$(SiH_3)_2O$	144.1°	163
NOCl	116°	114 (NO)
NO_2^-	115.4°	123.6
NO_2	134.15°	119.7

(c) bond angles between the limits 180° and 109° 28′ have varying amounts of π overlap; and (d) angles around 120° suggest that just one π bond is formed.

The bond angles in OCl_2, $HOCl$, $(CF_3)_2O$, $(CH_3)_2O$, and p-dimethoxy-benzene are greater than the tetrahedral angle 109° 28′, which is taken as evidence for some multiple bonding. The first two compounds may be represented as follows:

$$\delta^- \underset{Cl}{\overset{\delta^+ \ \ O}{C l}} \ \ \longleftrightarrow \ \ \underset{Cl \ \ \ \ Cl}{\overset{O}{\diagup}} \ \ \underset{\delta^-}{} \ \ \underset{H \ \ \ \ Cl}{\overset{\delta^+ \ \ O}{\diagup}} \ \delta^-$$

The π component in the O—Cl bond (and which cannot arise in an O—F bond) will increase both the oxygen s character and the electron density in the bond and lead to an increase in the bond angle. For the last three compounds mentioned, it is necessary to invoke a hyperconjugative mechanism to give some π component to the bonds. For example, in $(CH_3)_2O$:

$$\underset{CH_3 \ \ \ \ CH_3}{\overset{O}{\diagup \diagdown}} \ \longleftrightarrow \ \underset{H^-CH_2 \ \ \ \ CH_3}{\overset{O^+}{\diagup \diagdown}} \ \longleftrightarrow \ \underset{CH_3 \ \ \ \ CH_2H^-}{\overset{O^+}{\diagup \diagdown}}$$

The three species NO_2^+, NO_2, and NO_2^- provide a good series in which to investigate the influence of π bonding.[6,7] From molecular orbital theory, it is possible to see why both the shape and bond length change with the number of valence electrons (Table 9.18). The ion NO_2^+ has 2σ and 2π bonds, (compare the molecular orbital diagram, Fig. 9.4). The additional electron in NO_2 must go into the antibonding $2\pi_u$ orbital; however, by bending the molecule (which destroys some of the π bonding), the doubly degenerate $2\pi_u$ orbital splits into $2a_1$ and $2b_1$, the first of which interacts strongly with

Table 9.18 Some Properties of NO_2^+, NO_2, and NO_2^-

Property	NO_2^+	NO_2	NO_2^-
Bond angle	180°	134°	115.4°
Bond length (pm)	115.4	119.7	123.6
Valence electrons	16	17	18

the higher energy $4a_1$ orbital. This interaction strongly stabilizes the $2a_1$ orbital, and therefore the seventeenth valence electron in NO_2 is now in a more stable orbital. In the case of NO_2^-, the eighteenth valence electron will pair up with the odd electron in the stabilized $2a_1$ orbital, and the extent of π bonding is further reduced. The magnetic properties of the NO_2^- ion agree with this picture; linear NO_2^- would have a paramagnetic ground state $(2\pi_u)^2$, which is not observed.

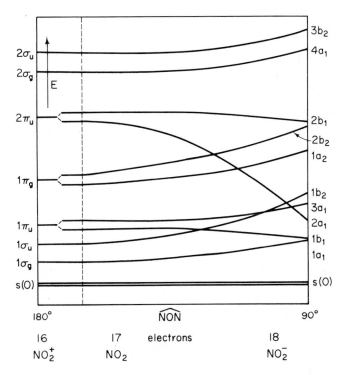

Fig. 9.4 Correlation diagram for the triatomic system NO_2^+, NO_2, and NO_2^-.

Angular Stereochemistry (Table 9.10)

The prediction of the configuration of maximum probability method, for a molecule such as water, is in good agreement with experiment (cf. Section 4.8). Since there are two lone pairs in AB_2 molecules (containing only single bonds) and only one in AB_3 molecules, we expect the bond angles to be smaller for the former species. However, when π bonding becomes significant, as in OCl_2 and $HOCl$, for example, this generalization is not true.

In a few cases, bond angles appear to be less than 90°, as in organic oxides such as ethylene oxide (Table 9.19). It is not energetically favorable to invoke the use of d orbitals to explain these small angles, and the best

Table 9.19　　Two-Coordinate Angular Compounds

	Compound	Bond lengths (pm)		Bond angles	
N	NH_2^-		103		104°
O	H_2O		95.7		104.52°
	CH_3-OH	CO	142.8	COH	109°
		OH	96.0		
	$(CH_3)_2O$		141.6		111.5°
	$(C_2H_5)_2O$		143		108°
	$(CF_3)_2O$		136		125°
	$HOCl$	OH	95.7		113°
		OCl	170		
	OCl_2		170		110.8°
	OF_2		141.3		103.8°
	NH_2-OH	OH	96	NOH	103°
		ON	146		
	SF_5-OH	SO	164	SOF	108°
		OF	143		
	H_2O_2 gas	OO	147	HOO	101°
		OH	97	Dihedral	106°
	H_2O_2 solid	OO	149	HOO	97°
				Dihedral	94°
	$(CH_3)_2O_2$	OO	144.5	COO	105°
	$(SF_5)_2O_2$	OO	147	SOO	105°
		SO	166	Dihedral	107°
	F_2O_2	OO	121.7	OOF	109° 30′
		OF	157.5	Dihedral	87° 30′
	$\begin{array}{c} CH_2 \\ \;\;\big\backslash \\ |\;\;\;\;O \\ \;\;\big/ \\ CH_2 \end{array}$ (ethylene oxide)	CO	143.6	COC	61.6°
	$\begin{array}{c} CH_2-O \\ |\;\;\;\;\;| \\ CH_2-CH_2 \end{array}$ (trimethylene oxide)	CO	146	COC	94.5°

Table 9.19 (*cont.*)

	Compound	Bond lengths (pm)		Bond angles	
	$CH_3-CH\overset{\diagdown \diagup}{\underset{O}{}}CH_2$ (propylene oxide)	CO	143.6	COC	58°
	$B_2O_5^{4-}(Co_2^{2+})$	BO	129, 130	BOB	140°
	$B_3O_6^{3-}(K_3^+)$		136	BOB	120°
	$(SiH_3)_2O$		163	SiOSi	144.1°
	P_4O_6		165		127.5°
	$Cr_2O_7^{2-}((NH_4^+)_2)$		191	CrOCr	115°
	$SbCl_5 \cdot POCl_3$	SbO	218	SbOP	143.7°
		PO	146		
	SeO_2		178		125°
	HgO		203		109°
	SiO_2 (α quartz)		161		142°
	CrO_3				136°
	PdO			PdOPd	98°
	O_3		127.8		116.8°
N	$(CH_3)_2C=NOH$	CN	129	CNO	111°
		N—OH	136		
	$CH_3-\underset{HON}{\overset{\|}{C}}-\underset{NOH}{\overset{\|}{C}}-CH_3$	CN	127	CNO	114°
		N—O	138		
	N_2F_2	NN	125	NNF	115°
		NF	144		
	$(CH_3)_2N_2$	NC	147	NNC	110°
		NN	124		
	NO_2		197		134° 15′
	NOF	NO	113	FNO	110°
		NF	152		
	NOCl	NO	114	ClNO	116°
		NCl	195		
	NOBr	NO	115	BrNO	117°
		NBr	214		
	HNO_2	NO	120	ONO	116°
		N—OH	146		
	NO_2^-	NO	123.6		115.4°
F	$(HF)_n$	HF	100		120°
		FH···F	255		
	MF_3(h.c.p.) Pd, Rh, Ir, Ru				132°
	MF_3 Intermediate packing Ti, V, Cr, Fe, Co				~150°
	MF_3 Cubic (ReO$_3$ structure) Mo, Ta, Nb				180°
	MF_5 Tetrameric Nb, Ta, Mo				180°
	SiF_4 Solid			Si—F···Si	180°
	$[(Et_3Al)_2F]K$			AlFAl	180°

explanations are either in terms of bent σ bonds or by the use of molecular orbital three-center bonds (Fig. 9.5). Obviously, these are special cases, and the bond angles must be determined by the three-membered ring geometry rather than by any subtle bonding requirements.

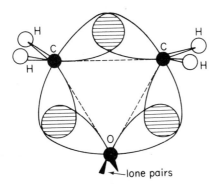

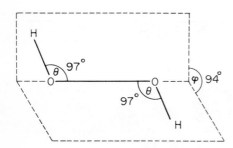

←lone pairs   **Fig. 9.5**   Bent σ bonds in ethylene oxide.

The oxygen atoms in peroxides R_2O_2 have an angular stereochemistry; in addition, the R groups are staggered (Fig. 9.6). It is therefore necessary to quote the dihedral angle when describing such molecules. We shall consider the reasons for the stereochemistry of these molecules using H_2O_2 as an

Fig. 9.6   Shape of H_2O_2.

example. The molecular orbitals and electron configuration for the peroxide ion O_2^{2-} are

$$\sigma(2s^b)^2\sigma(2s^*)^2\sigma(2p_z^b)^2\pi(2p_{xy})^4\pi(2p_{xy}^*)^4$$

(based on molecular oxygen and ignoring the 1s orbital). Hence, the O—O bond is a single $\sigma(2p_z^b)^2$ bond. Since the π bonding is effectively removed by the π^* electrons, we shall assume that the p_x and p_y orbitals remain non-bonding and localized on each oxygen atom (Fig. 9.7). If the first proton (H^+) bonds at $p_x(a)$ on oxygen a, the second proton can bond at three positions: $p_x(b)$; $p_y(b)$; or $\sigma(2s^*)$. If it bonds at $p_x(b)$ or $\sigma(2s^*)$, strong non-

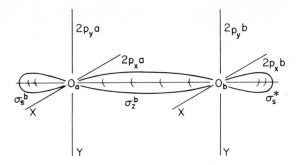

Fig. 9.7 Bonding and nonbonding orbitals of O_2^{2-}.

bonding electron repulsions will arise from the filled free p_y orbitals. The second proton therefore bonds with $p_y(b)$, leaving noninteracting lone pairs $[\sigma(2s^b)^2, \sigma(2s*)^2]$ and weakly interacting lone pairs $[(2p_v)_a^2, (2p_x)_b^2]$ on each oxygen. The staggered configuration is thus preferred, and the angles θ and φ are expected to be in the vicinity of from 90 to 100°, depending on the amount of s and p character in the various bonds. The change in θ and φ with change in the physical state of H_2O_2 (Table 9.19) is probably associated with varia-tions in intermolecular hydrogen bonding.

A molecule with four valence electrons less than H_2O_2, e.g., acetylene C_2H_2, will have no filled antibonding π orbitals, and the stable arrangement, which retains the π bonding, is a linear molecule.[8] A 12-valence electron molecule, such as N_2F_2 (excluding the fluorine nonbonding electrons) will be intermediate in structure between C_2H_2 and H_2O_2, so that one π bond will contribute to the N—N bond, giving a planar molecule with three possible isomeric forms (Fig. 9.8). The second form (b) is expected to be the more stable, and both the cis and trans forms of N_2F_2 have been isolated.

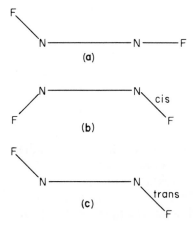

Fig. 9.8 Isomers of N_2F_2.

Organic oxime- and azo-compounds also exhibit stereoisomerism,

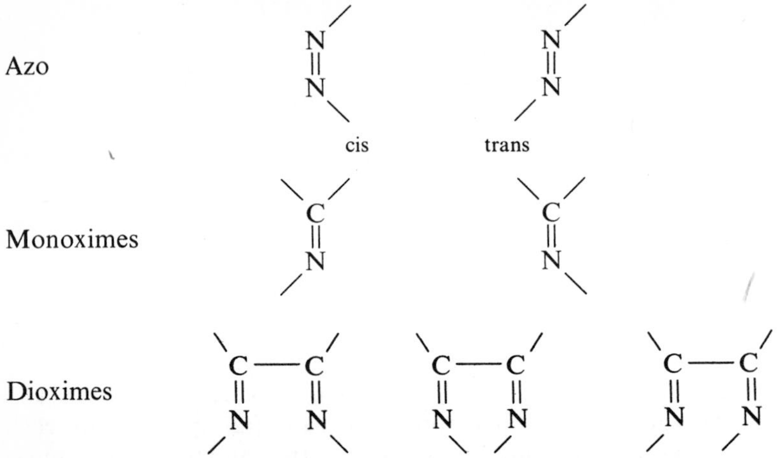

Azo

cis   trans

Monoximes

Dioximes

Evidence for all isomeric forms has been found. The lone pair in acetoxime is involved in the H-bond $N \cdots H—O$, and this is a guide to its position (Fig. 9.9).

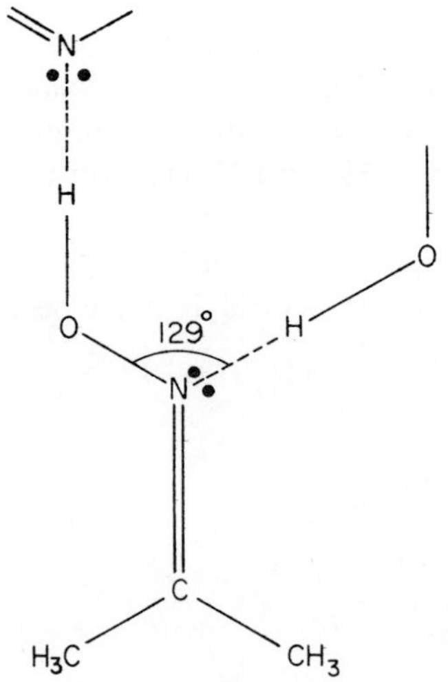

Fig. 9.9 Acetoxime.

The stereochemistry of two-coordinate fluorine is either angular (bond angles 120–140°) or linear. A linear fluorine bridge is probably determined by (a) the stereochemical requirements of the atoms it bridges, and (b) the formation of a $F \rightarrow M$ π bond. The Al—F bonds in the linear fluorine bridge in $[Et_3Al\text{-}F\text{-}AlEt_3]^-$, for example, probably involve a π

component. A similar situation is likely for the early transition metal fluorides, e.g., $(MoF_5)_4$, in high oxidation states. The bent fluorine bridge is the normally expected shape and occurs where the stereochemical requirement of the fluorine dominates. The structure of polymeric hydrogen fluoride demonstrates this (Fig. 9.10).

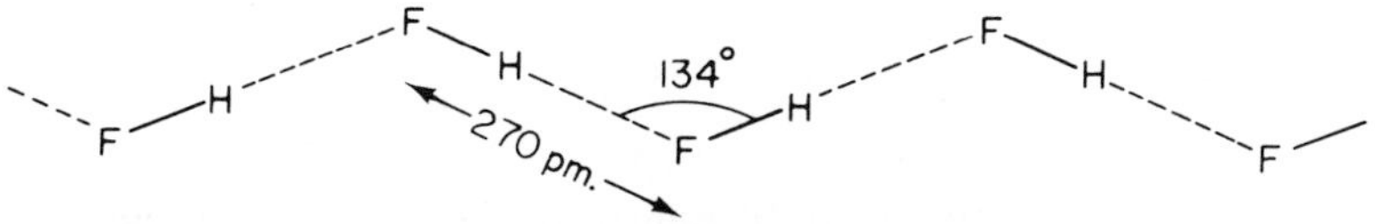

Fig. 9.10 $(HF)_n$.

Linear Stereochemistry (Table 9.20)

Linear molecules occur for two-coordinate Be, C, N, and O, where the bonding consists in two σ bonds and (especially for C, N, and O) two π bonds. The two π bonds can be either on one side of the central atom (as in HCN) or one on each side (as in CO_2).

The C—X (X = Cl, Br, I) bonds in the compounds $RC{\equiv}CX$ and $N{\equiv}CX$ are short in comparison with the same bond in CX_4 molecules

Table 9.20 Two-Coordinate Linear Compounds

	Compound		Bond length (pm)		
Be	$BeCl_2$ (vapor)		117		
	BeI_2 (vapor)		212		
C	CO_2		116.32		
	CS_2		155.32		
	COS	CO	116.37		
		CS	155.86		
	HNCO	NC	118.3		
		CO	118.4		
	HNCS	CS	156.1		
		CN	121		
	$RC{\equiv}CR$	CC	120.8		
N	$X{-}CN$	C—Cl	157	CN	116
		C—Br	179	CN	115.8
		C—I	199.5		
	N_3^-		115		
	N_2O	NN	112.6		
		NO	118.6		
O	$[MCl_5{-}O{-}MCl_5]^{4-}$	Ru—O	180		
	Re, Ru	Re—O	186		

Table 9.21   Carbon–Halogen Bond Lengths

	C—X Single bond in CX_4 (pm)	C—X in acetylenes (pm)	C—X in cyanides (pm)
Cl	177	164	157
Br	194	179	179
I	215	199	199.5

(Table 9.21). The shortening can be related to the greater s character of the σ bonds compared with σ bonds in a tetrahedral molecule, and is evidence in favor of the use of sp hybrids by the carbon atom in the linear molecules.

The linear oxygen bridge in compounds such as $K_4[Ru_2OCl_{10}]$ and $K_4[Re_2OCl_{10}]$ strongly suggests multiple bonding between the oxygen and metal ions, which is supported by the observed diamagnetism of these compounds.

9.7   ELECTRON-DEFICIENT COMPOUNDS[9,10]

Electron-deficient compounds have gained attention throughout the years because their existence could not be described in terms of early bonding theories. However, they are readily described by the molecular orbital theory using three-center bonds. It is common practice in MO theory to use two-electron n-center orbitals, and two-center bonding is a special case of the more general situation.

Electron deficiency occurs when the number of available valence electrons is less than the number of bonding orbitals. It therefore occurs especially for the elements of Groups I, II, and III. The compounds are generally formed with hydrogen and alkyl ligands, and some chlorides are also known, e.g., B_4Cl_4 and B_8Cl_8. Boron in BF_3 is theoretically electron deficient, but, as discussed earlier, the deficiency is overcome by the formation of a $B \leftarrow F$ π bond. When the ligand cannot form a π bond, as is the case for hydrogen and alkyl groups, then stable polymeric compounds are formed. Exceptions to this behavior are known; e.g., BMe_3 is monomeric. In this case, hyperconjugation has been invoked to allow some π-type interaction to remove the electron deficiency on the boron.

A three-center bond is formed by the overlap of three bonding orbitals giving rise to three molecular orbitals. The overlap can take place in two ways, either "open" or "closed." Both the "open" three-center bond [Fig. 9.11(a), (b)] and the "closed" three-center bond (Fig. 9.12) occur in the boron hydrides, and the latter is probably the type that exists in the

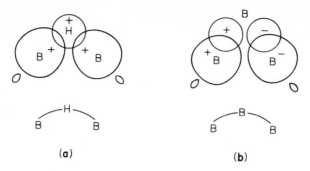

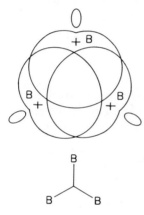

Fig. 9.11 Open three-center bonds in the boron hydrides.

Fig. 9.12 Closed three-center bond in the boron hydrides.

alkyl derivatives of the Group I, II, and III elements. A three-center molecular orbital diagram is given in Fig. 9.13. Here, the bonding orbital ψ_1 will hold two electrons, and an additional two electrons $(\psi_2)^2$ will not destroy the bonding; however, the configuration $(\psi_1)^2(\psi_2)^2$ will probably be less favorable than two two-center bonds. Thus, three-center overlap gives three molecular orbitals, which require just two electrons for bonding. In this way, the electron-deficient nature of compounds such as $[(C_2H_5)Li]_4$, $(InH_3)_4$, B_2H_6, and $(AlMe_3)_2$ is overcome. The coordination number of the bridging atom in such compounds is increased, so that the bridging carbon atoms in $(AlMe_3)_2$, for example, are five coordinate.

It is possible to calculate the H_bBH_b angle in B_2H_6, knowing that the HBH angle is 121.5° (Fig. 9.14) and using the criterion of maximum overlap. The best boron orbitals to bond the terminal hydrogen atoms at an angle of 121.5° are close to sp^2 hybrids (angle 120°). The orbitals remaining on the boron available for the three-center bond are one sp^2 hybrid and one p

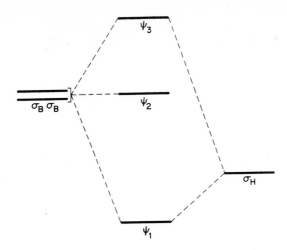

Fig. 9.13 Molecular orbital energy diagram for a three-center bond.

orbital (p_z). With the arrangement of axes shown in Fig. 9.14, the sp^2 orbital is $(1/\sqrt{3})s + (2/\sqrt{6})p_x$. Two new equivalent boron orbitals can be formed from this sp^2 hybrid orbital, and the unused p_z orbital and these are

$$\psi_1 = \frac{1}{\sqrt{6}}s + \frac{1}{\sqrt{3}}p_x + \frac{1}{\sqrt{2}}p_z$$

$$\psi_2 = \frac{1}{\sqrt{6}}s + \frac{1}{\sqrt{3}}p_x - \frac{1}{\sqrt{2}}p_z$$

Substitution of the analytical forms of the s, p_x, and p_z orbitals (angular function only) into ψ_1 gives

$$\psi_1 = \frac{1}{\sqrt{6}} + \sin\theta\cos\varphi + \sqrt{\frac{3}{2}}\cos\theta$$

In the xy plane, $\varphi = 0$; therefore,

$$\psi_1 = \frac{1}{\sqrt{6}} + \sin\theta - \sqrt{\frac{3}{2}}\cos\theta$$

The first derivative $d\psi_1/d\theta$, when equated to zero, will allow the value of θ to be obtained corresponding to the orbital giving maximum overlap in the three-center bond.

$$\frac{d\psi_1}{d\theta} = \cos\theta - \sqrt{\frac{3}{2}}\sin\theta = 0$$

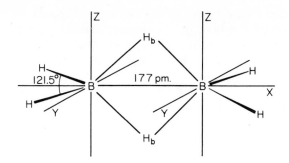

Fig. 9.14 Coordinate system for B_2H_6.

Therefore,

$$\tan \theta = \sqrt{\frac{2}{3}}$$

$$= 39° \, 14'$$

and the bond angle

$$H_bBH_b = (180 - 2\theta)$$

$$= 101° \, 32'$$

This figure agrees reasonably well with the experimental figure of 97°.

The majority of the structures of the boron hydrides are based on the icosahedral polyhedra (Fig. 9.15). An icosahedron consists of 20 triangular faces meeting in 12 corners, five triangles meeting at each corner. Elementary boron, B_{12} (α–rhombohedral form), crystallizes with this structure. Each

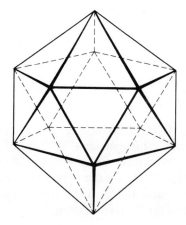

Fig. 9.15 Icosahedron.

boron is bonded to five others within the icosahedron (B—B, 173–179 pm). Half the boron atoms bond by two-center orbitals with boron atoms of neighboring icosahedrons (B—B, 171 pm). The remaining six bond by three-center orbitals to two boron atoms, one from each of two neighboring icosahedrons (B—B, 203 pm).

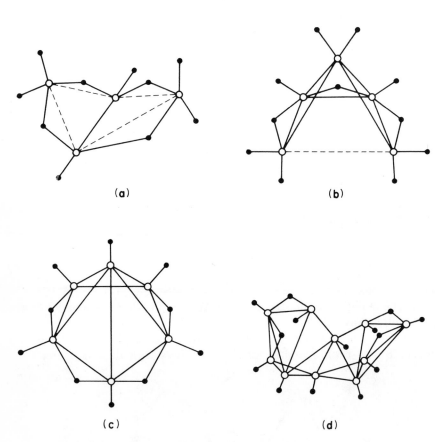

Fig. 9.16   Boron hydrides, icosahedral fragments: (a) B_4H_{10}, (b) B_5H_{11}, (c) B_6H_{10}, (d) $B_{10}H_{14}$.

The hydrides B_2H_6, B_4H_{10}, B_5H_{11}, B_6H_{10}, B_8H_{12}, B_9H_{15}, and $B_{10}H_{14}$ can be considered as icosahedral fragments (Fig. 9.16). The structures are therefore pseudoelemental in nature—i.e., fragments of elemental boron. Therefore, hydrides of the type B_7H_x are not expected to exist. The boron hydride ions, such as $B_{10}H_{10}^{2-}$ and $B_3H_8^-$, and carboranes, such as $B_{10}C_2H_{12}$, also have structures based on the icosahedron.

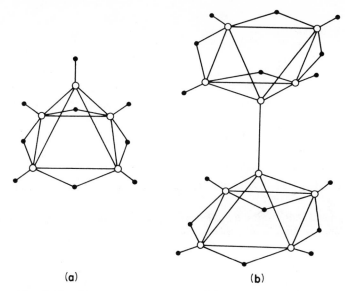

(a)   (b)

Fig. 9.17   Boron hydrides, octahedral fragments: (a) B_5H_9, (b) $B_{10}H_{16}$.

The hydrides B_5H_9 and $B_{10}H_{16}$ are exceptions to the preceding; B_5H_9 is a fragment of an octahedron (square-based pyramid), whereas $B_{10}H_{16}$ is two such fragments joined together (Fig. 9.17).

BIBLIOGRAPHY

1. R. J. GILLESPIE and R. S. NYHOLM, *Quart. Rev. London*, **11**, 339 (1957).

2. Structural parameters in Table 9.9 and following tables are obtained from L. E. SUTTON, ed., *Interatomic Distances, Chem. Soc. London Spec. 11* (1958), *Suppl. Spec. Publ. 18* (1965); and A. F. WELLS, *Structural Inorganic Chemistry*, 3rd ed., Clarendon Press, Oxford, Engl., 1962. More recent data are obtained from the literature.

3. H. A. BENT, *J. Chem. Educ.*, **37**, 616 (1960).

4. H. A. BENT, *Chem. Rev.*, **61**, 275 (1961).

5. F. A. COTTON, *J. Chem. Phys.*, **30**, 993 (1959).

6. C. J. BALLHAUSEN and H. B. GRAY, *Molecular Orbital Theory*, W. A. Benjamin, Inc., New York, 1965.

7. H. B. GRAY, *Electrons and Chemical Bonding*, W. A. Benjamin, Inc., New York, 1964.

8. A. D. WALSH, *J. Chem. Soc.*, 2288 (1953).

9. E. L. MUETTERTIES, ed., *The Chemistry of Boron and Its Compounds*, John Wiley & Sons, Inc., New York, 1967.

10. W. N. LIPSCOMB, *Boron Hydrides*, W. A. Benjamin, Inc., New York, 1963.

FURTHER READING

B. E. DOUGLAS and D. H. MCDANIEL, *Concepts and Models in Inorganic Chemistry*, Blaisdell Publishing Co., New York, 1965.

F. A. COTTON and G. WILKINSON, *Advanced Inorganic Chemistry*, 3rd ed., Interscience Publishers, New York, 1972.

J. N. MURRELL, S. F. A. KETTLE, and J. M. TEDDER, *Valence Theory*, 2nd ed., John Wiley & Sons, Inc., New York, 1970.

D. J. ROYER, *Bonding Theory*, McGraw-Hill Book Company, New York, 1968.

M. C. DAY and J. SELBIN, *Theoretical Inorganic Chemistry*, 2nd ed., Reinhold Publishing Corp., New York, 1969.

K. B. HARVEY and G. B. PORTER, *Introduction to Physical Inorganic Chemistry*, Addison-Wesley Publishing Co., Inc., Reading, Mass., 1963.

10

The Stereochemistry of
the *p*-Block Elements

The elements to be considered in this chapter are characterized by partially filled np valence orbitals, where $n = 3$ to 6 (Table 10.1). They are a diverse group of elements but have some features in common, such as (a) access to vacant nd orbitals, allowing for expansion of the valency shell, and (b) an increase in the stability of low oxidation states as n increases.

10.1 COORDINATION NUMBER

The coordination numbers displayed by the elements are listed in Table 10.2. The details in the table are based largely on available structural information and may exclude coordination numbers inferred from other chemical evidence. Several facts stand out from this table: (1) only a few elements have coordination numbers greater than six, and these are the elements with the larger atomic radii; (2) the oxidation state of an element is important in determining the coordination number it displays; and (3) the dominant coordination numbers are four and six. Points 1 and 2 will be discussed in more detail.

Table 10.1 The *p*-Block Elements

n	np^1	np^2	np^3	np^4	np^5	np^6
3	Al	Si	P	S	Cl	
4	Ga	Ge	As	Se	Br	
5	In	Sn	Sb	Te	I	Xe
6	Tl	Pb	Bi	Po		

Table 10.2 The Coordination Number of the *p*-Block Elements*

Coordination Number	Group 2	3	4	5	6	7	8
					S(0)		
2	Zn		Sn(II)		S(II)	Cl	Xe
	Cd		Pb(II)		Se(II)	Br	
	Hg				Te(II)	I	
3	Hg	Al(III)	Ge(II)	P(III)	S	Cl	Xe
		Ga(III)	Sn(II)	As(III)	Se	Br	
		In(III)	Pb(II)	Sb(III)	Te	I	
		Tl(III)		Bi(III)			
4			Si(0)				
	Zn	Al	Si(IV)	P	S	Cl	Xe
	Cd	Ga	Ge(IV)	As	Se(IV)	Br	
	Hg	In	Sn(IV)	Sb	Te(IV)	I	
		Tl	Pb(IV)				
5	Zn	Al	Si	P(V)	S	Br	Xe
	Hg	In	Sn	As(V)	Se	I	
		Tl		Sb(V)	Te		
				Bi(V)			
6	Zn	Al	Si	P	S	I	Xe
	Cd	Ga	Ge	As	Se		
	Hg	In	Sn	Sb	Te		
		Tl	Pb	Bi			
7			Pb			I	Xe
8	Cd	Tl	Sn				Xe
			Pb				
9			Sn				
			Pb				

* Principal coordination numbers for the elements are indicated by underlining.

Coordination Number and Atomic Size

The atomic size increases down a periodic group and the coordination number may be expected to increase. This is demonstrated from the data given in Table 10.3. There is a trend of increase in the maximum coordination number down a group, but with a marked change occurring from the first to the second short period. However, if one considers the same ligand, such as oxygen or chlorine (columns 4 and 5, Table 10.3), a distinct trend is apparent.

Since the elements (other than those of the first short period) have accessible vacant d orbitals, it is easy to see why they can achieve coordination

Table 10.3 Coordination Number of the *p*-Block Elements

Element	Covalent radius (single bond) (pm)	Maximum coordination number	Coordination number for oxyanions	Coordination number for chloro-compounds	
B	81	4	3, 4	4	
Al	123	6	6	4	MCl_4^-
Ga	125	6	6	4	
In	150	6	6	6	MCl_6^{3-}
Tl	155	8	6	6	
C	77	4	2	4	MCl_4
Si	117	6	4(6)	4	
Ge	122	6	6	6	
Sn	140	9	6	6	MCl_6^{2-}
Pb	154	9	6	6	
N	70	4	3	3	MCl_3
P	110	6	4	6	
As	121	6	4	6	MCl_6^-
Sb	141	6	6	6	
Bi		6	6	6	
O	66	4	2	2	OCl_2
S	104	6	4	4	SCl_4
Se	117	6	4	6	MCl_6^{2-}
Te	137	6	6	6	
F	64	4	1		
Cl	99	4	4		
Br	114	5	4		
I	133	7	6		
Xe		8	6	8	XeF_8^{2-}
Zn	125	6		4	$ZnCl_4^{2-}$
Cd	141	8		6	$CdCl_6^{4-}$
Hg	144	6		4	$HgCl_4^{2-}$

numbers greater than four. However, these nd orbitals tend to be diffuse and inefficient in bonding, particularly as n increases. Hence, we find that whereas increase in atomic size can give rise to higher coordination numbers, the fall-off in efficient orbital overlap may offset this. This may explain why the coordination numbers of the elements tend to level out at six. The formation of the fluoroanion SiF_6^{2-}, but not the corresponding chloroanion $SiCl_6^{2-}$, is explained by the diffuse nature of the d orbitals. The more electronegative fluorine withdraws electron density from the silicon atom, increasing its effective nuclear charge. This causes a contraction of the d orbitals, making them more suitable for bonding. This effect occurs to a lesser extent for a chloro- complex, and therefore such anions occur mainly for the larger atoms.

Coordination Number and Oxidation State

The heavier members of the p-block elements display an increased stabilization of the lower valency states—i.e., In(I), Tl(I); Sn(II), Pb(II); Sb(III), Bi(III); and Se(II), Te(II), compared with the usual oxidation levels III, IV, V, and IV, respectively. Decreasing bond strength down a group, for covalent compounds, is less likely to compensate for the promotion energy [e.g., M(II) to M(IV), which increase down a group], and this would favor the lower oxidation state.[1] In the case of ionic compounds, lattice energies decrease down a group and may not compensate for the high ionization energies involved (e.g., $M^{2+} \rightarrow M^{4+} + 2e$); this again favors the lower oxidation state.[1] The elements in the low oxidation state have a lone pair of electrons [two lone pairs for Se(II) and Te(II)] called the "inert pair." However, the lone pair is not normally stereochemically "inert," and it is not necessarily chemically inert because it can be used by donation in chemical bonds.

It is expected that the presence of the lone pair in the low oxidation states would result in lower coordination numbers compared with the elements in their higher oxidation states. The greater positive charge for the elements in the higher oxidation levels will assist in this difference by causing d-orbital contraction, and by virtue of the electroneutrality principle. However, as seen from Table 10.4, there does not seem to be any definite correlation between valency state and the maximum coordination number displayed by the elements. Clearly, however, from Table 10.5, the coordination number of elements in the lower valency states, in compounds in the gaseous or liquid state (such as fluorides and chlorides), is less than for the same elements in their higher oxidation states. Since compounds of the elements in the lower oxidation state tend to be ionic in character, the same

Table 10.4   Coordination Number and Oxidation State

Low oxidation level	Maximum coordination number	Compound	High oxidation level	Maximum coordination number	Compound
Ga(I)	4?	GaL_4^+	Ga(III)	6	GaF_6^{3-}
In(I)	5	InBr	In(III)	6	$[InCl_5H_2O]^{2-}$
Tl(I)	8	$TlNO_3(thiourea)_4$	Tl(III)	6	$TlCl_6^{3-}$
Ge(II)	6	GeI_2	Ge(IV)	6	$GeCl_6^{2-}$
Sn(II)	9	$SnCl_2(c)*$	Sn(IV)	6	SnF_6^{2-}
Pb(II)	9	$PbF_2(c)$	Pb(IV)	6	$PbCl_6^{2-}$
P(III)	4	PH_4^+	P(V)	6	PCl_6^-
As(III)	6	AsI_3	As(V)	6	AsF_6^-
Sb(III)	6	$SbCl_6^{3-}$	Sb(V)	6	$SbCl_5 \cdot POCl_3$
Bi(III)	6	$BiCl_6^{3-}$	Bi(V)	5	Ph_3BiCl_2
S(II)	2	SCl_2	S(IV)	4	SCl_4
Se(II)	2?	$Se(CH_3)_2$	Se(IV)	6	$SeCl_6^{2-}$
Te(II)	4	$TeCl_2$ (thiourea)	Te(IV)	6	$TeCl_6^{2-}$
			S(VI)	6	SF_6
			Se(VI)	6	SeF_6
			Te(VI)	6	TeO_6^{6-}

* (c) = Crystalline.

Table 10.5   Coordination Number, Oxidation State, and Physical State

Element	Oxidation state	Halide	Coordination number* g, l	Coordination number* s	Oxidation state	Halide	Coordination number* g, l	Coordination number* s
Ga	I	GaF	1		III	GaF_3	3	6
In	I	InCl	1		III	$InCl_3$	4	
Tl	I	TlCl	1	6	III	$TlCl_3$		6
Ge	II	GeF_2		3	IV	GeF_4	4	
Sn	II	$SnCl_2$	2	9	IV	$SnCl_4$	4	
Pb	II	$PbCl_2$	2	9	IV	$PbCl_4$	4	
P	III	PCl_3	3		V	PCl_5	5	4, 6
As	III	$AsCl_3$	3		V	AsF_5	5	
Sb	III	$SbCl_3$	3	3	V	$SbCl_5$	5	5
Bi	III	$BiCl_3$	3	6 in BiF_3				
S	II	SCl_2	2		IV	SF_4	4	4
Se	II	$SeCl_2$	2		IV	$SeCl_4$	4	3?
Te	II	$TeCl_2$	2		IV	$TeCl_4$	4	3?

* g = gas, l = liquid, s = solid.

trend does not hold for the crystalline compounds. It is generally found that an increase in the covalent character of bonds tends to lower the coordination number, inasmuch as the limiting factor is available bonding orbitals and not the packing of ions around each other as in ionic compounds. Hence, the interplay of such factors as atomic size, charge, valency state, ionic and covalent bonding may well be the reason for the lack of definitive trends in the coordination numbers displayed by the *p*-block elements.

Coordination Numbers and Next-Nearest Neighbors

It is sometimes difficult to be sure of the exact coordination number of an atom, because the bonded ligands, or ions, can be either nearest or next-nearest neighbors. For example, the coordination number of tin in crystalline $SnCl_2$ can be taken as nine, but the individual bond lengths (in pm) are 266, 2 of 278, 2 of 306, 322, 330, and 2 of 386—a range of 120 pm. It becomes necessary to determine the maximum distance that two atoms can be apart and still be considered as involved in each other's coordination sphere. It often helps to consider the stereochemistry; for example, for the compound As[NN-diethyldithiocarbamate]$_3$ the dithio-carbamate ligand is a bidentate, but one sulfur of each ligand is at 235 pm from the arsenic whereas the other is at 280 to 290 pm. Hence, one can express the coordination number as three (pyramidal shape) or six (distorted trigonal antiprism).

10.2 STEREOCHEMISTRY

The stereochemistry of the *p*-block elements is summarized in Tables 10.6, 10.7, and 10.8. In Table 10.6, the arrangement of the various electron pairs is considered for each coordination number. The same information is provided in Table 10.7 in a different manner. Finally, in Table 10.8, a pictorial representation of the relationship between the various shapes is given (no differentiation is made between single and double bonds in this table).

10.3 TWO COORDINATION

Linear Molecules

Only a few, linear, two-coordinate molecules or ions occur for the *p*-block elements; these are principally found in the interhalogen compounds of bromine and iodine, where there are three lone pairs in

Table 10.6 Stereochemistry of the *p*-Block Elements

Shape	σ Bonds	Lone pairs	π Bonds	Elements
Two Coordination				
Linear	2	0	1 or 2 possible	Tl, Zn, Cd, Hg, Te?
	2	3	0	I, Xe, Br
Angular	2	1	Possible	Sn(II), Pb(II), S(db), Se?
	2	2	0	S, Se, Te
	2	2	1 or 2 possible	Cl, Br, I
Three Coordination				
Planar	3	0	1 or more possible	S, Al, Ga, In, Tl? Hg
T-shape	3	2	0	Cl, Br, I
Pyramidal	3	1	0	P, As, Sb, Bi, Ge(II), Sn(II), Pb(II), Tl(I)
	3	1	1	S, Se, Te
	3	1	2 or 3	Xe, Cl, Br, I
	3	1	Possible	Tl?
Four Coordination				
Tetrahedral	4	0	0	Al, Ga, In, Tl, Si, Ge, Sn, Pb, Zn, Cd, Hg, P, As, Sb
	4	0	1 or more	P, As, Sb, Cl, I, S, Se, Te?
Distorted tetrahedron	4	1	0	S, Se, Te, I, Sb, As, Ga(I)
	4	1	?	Sn, Pb
Square plane	4	2	0	Br, I, Xe, Te(II)
	4	0	?	Hg? Pb?
Five Coordination				
Trigonal bipyramid	5	0	0	Al, S, P, As, Sb, Bi, In, Sn, Zn, Cd
Square pyramid	5	1?	0	In, Tl, Zn? Hg
	5	1	0	Sb(III), Bi(III), Br, I
	5	1	1	Se, I
Six Coordination				
Octahedron	6	0	0	Al, Ga, In, Tl, Si, Ge, Sn, Pb, P, As, Sb, Bi, S, Zn, Cd, Hg
	6	1	0	Se, Te, Xe, I?
	6	0	Double bonds	I

addition to the two bonding pairs. The compounds are of the type Br_3^-, ICl_2^-, and I_5^-; xenon difluoride is another example (Table 10.9). The bond lengths are long but are consistent with a trigonal bipyramidal arrangement of electron pairs with the polar positions corresponding to the bonds. The effect of neighboring cations is said to be the cause of the difference in the bond lengths on either side of the central atom in certain anions, such as in I_3^-.

Table 10.7   Stereochemistry of the *p*-Block Elements

	Linear		Angular		Planar		Pyramidal		Tetrahedral		Distorted Tetrahedral		Square Planar		Trigonal Bipyramid		Square Pyramid		Octahedron					
π	1 or 2	0	poss	0	1 or 2	poss	0	0	1	2 or 3	poss	0	1 or more	0	?	0	?	0	0	0	1	0	0	poss	
lone	0	3	1	2	2	2	0	2	1	1	1	1	0	0	1	1	2	0	0	1?	1	1	0	0	1	0
σ	2	2	2	2	2	2	3	3	3	3	3	3	4	4	4	4	4	4	5	5	5	5	5	6	6	6	7	8	9
Al																													Al
Ga																													Ga
In																													In
Tl					?		I			?																			Tl
Si																													Si
Ge							II																						Ge
Sn		II					II																			II	II		Sn
Pb		II					II							?											II	II	II		Pb
P																													P
As																													As
Sb																			III										Sb
Bi																			III										Bi
S																													S
Se			?																										Se
Te															?		II												Te
Cl																													Cl
Br																													Br
I																						?							I
Xe																													Xe
Zn																													Zn
Cd																													Cd
Hg														?												?			Hg

poss = possible

▮ = where compounds are known with the stereochemistry indicated.

? = some doubt over the stereochemistry.

II = oxidation state of element; otherwise the "normal" oxidation state is assumed.

Linear molecules are also formed by Zn(II), Cd(II), and Hg(II), which, although not strictly *p*-block elements, will be discussed here. The linear compounds occur most frequently for mercury,[2,3] which has been attributed to the large 6*s*-6*p* separation of about 2 eV for this element. For a small *ns*-*np* separation, a coordination number of four may be expected, but as the

Table 10.8   Stereochemistry of the *p*-Block Elements—Two Coordination to Six Coordination

Octahedron	Sq. based pyramid	Square plane	T shape	Linear
Al, Ga, In, Tl, Si, Ge, Sn, Pb, P, As, Sb, Bi, S, Se, Te, Zn, Cd, Hg	Se(db), Sb, Bi, Br, I, In, Tl, Zn?, Hg	Br, I, Xe Te(II), Hg (no lone pairs), Pb?		

Trigonal bipyramid	Distorted tetrahedron	T shape	Linear
Al, In, P, As, Sb, Bi, S, Sn, Zn, Cd	S, Se, Te, I, Sb, As, Ga(I), Sn, Pb	Cl, Br, I	Br, I, Xe

Tetrahedron	Pyramidal	Angular
Al, Ga, In, Tl, Si, Ge, Sn, Pb, P, As, Sb, Zn, Cd, Hg, Cl, I, S, Se, Te?	P, As, Sb, Bi, Ge(II), Sn(II), Pb(II), Tl(I), S, Se, Te, Cl, Br, I, Xe, Tl?	S, Se, Te, Cl, Br, I

Trigonal plane	Angular
Al, Ga, In, Tl? S, Hg	Sn(II), Pb(II), S, Se?

Linear
Tl, Zn, Cd, Hg, Te?

◗ = Lone pair

Table 10.9[4,5]   Stereochemistry of Interhalogen Species

Compound or ion		Bond length (pm)	Bond angle		
Br_3^-		253, 254	171°		Br—Br in Br_2 is 228.4 pm
$IBrCl^-$	ICl	238	linear		I—Cl in
ICl_2^-		255 (Me_4N^+)	linear		ICl_4^- is 232.1 pm
		269, 247			
		$(C_4H_{10}N_2H_2^{2+})$			
pyICl	ICl	251	NICl	180°	
I_3^-		283, 303	linear		I—I in
		290			I_2 is 266.6 pm
I_5^-		314, 293	176°,	175°	
		317, 281			
I_8^{2-}		284, 300	176°	175°	
$(CH_3)_3NI_2$	I—I	283	NII	179°	
BrI_2^-	I—I	278	BrII	178°	
XeF_2		200	180°		Xe—F in
					XeF_4 is 195 pm

energy separation increases, the coordination number will drop to three or two. This argument implies covalent bonding using $nsnp^x (x = 1, 2,$ or 3) hybrids. If the bonding also involves some d_{z^2} component (i.e., a $d_{z^2} s$ hybrid), then the $(n - 1) d - ns$ energy separation is now relevant. The energies are 9.7 eV (Zn^{2+}), 10.0 eV (Cd^{2+}), and 5.3 eV (Hg^{2+}). The low value for mercury also indicates that this metal will favor a two-coordinate linear stereochemistry over the other Group II metal ions.

The linear cation $[CH_3\text{-}Tl\text{-}CH_3]^+$ of trivalent thallium is isoelectronic with $(CH_3)_2Hg$; therefore, the linear stereochemistry is not unexpected.

Bent Molecules

Bent triatomic molecules occur when two of four σ pairs are bonding, as in H_2S, or when two of three σ pairs are bonding. In the latter case, π bonding is possible.

The element that forms the greatest number of triatomic bent molecules is sulfur. In most cases, the sulfur bond angles lie within the range of from 90 to 110° (Table 10.10). Unlike oxygen, bond angles greater than 110° are not observed, indicative of a sharp decrease in the multiple-bonding ability of sulfur compared with that of oxygen. Except for $TeCl_2$, whose structure

Table 10.10 Stereochemistry of *MXX'* Compounds of Sulfur, Selenium, and Tellurium

XX'	S	Se	Te
H, H	92.2°	91°	89.5°
Cl, Cl	103°		150–180°?
CCl_3, Cl	109°		
CH_3, H	100.3°		
CF_3, CF_3	105.6°	104.4°	
CH_3, CH_3	105°	98°	
CN, CN		119°	
Br, Br			98°
SiH_3, SiH_3	100°		

needs confirming, and where comparisons are possible, the bond angles for *MRR'* compounds (M = S, Se, and Te) decrease as the central atom gets bigger.

The same range of bond angles is observed in molecules of the type M_2R_2 (Table 10.11). This also applies to organic compounds, where the Group VI element is a heteroatom, except in the case of ethylene sulfide, where the sulfur bond angle is 65.8°, analogous to the oxygen bond angle in ethylene oxide (p. 150).

The cations BrF_2^+ and ICl_2^+ are angular. The ICl_2^+ cation is present in the compound $SbCl_8I$, which is best represented as $ICl_2^+SbCl_6^-$. Each iodine is surrounded by two chlorine atoms at 229 and 231 pm (bond angle 91.5°). Two more chlorines are at a distance of 294 pm, suggesting some weak interaction. The I—Cl bonds in ICl_2^+ are normal length and not lengthened as in linear ICl_2^-.

A few examples are known of compounds with only three σ pairs (cf. end of Table 10.11). In addition, a vacant p orbital (or d orbital) may take part in multiple bonding, as in SO_2. The same applies for the gaseous dihalides of Sn(II) and Pb(II), especially for the least electronegative halogens. The Sn, Pb—X bonds are shorter in the gaseous state than in the solid state, where the coordination number is greater. This may be due to some π bonding in the gaseous compounds, although the change from discrete molecular to mainly ionic lattice bonding may also be the reason.

The alkaline earth metals are not members of the *p*-block elements, but it is appropriate to consider the shape of the dihalides of these metals at this point. The halides listed at the end of Table 10.11 are bent, whereas the rest are linear. The ligand arrangement around the metal atom has been deduced from molecular beam deflection experiments and from vibrational spectra of the difluorides trapped in krypton at 20 K. The nonlinear shape

Table 10.11   Stereochemistry of M_2R_2 and MR_2 Compounds

Compound	Bond length (pm)		Bond angle	
$(CF_3)_2S_2$	CS	182.9	CSS	105.4°
	SS	205.3		
$(CF_3)_2Se_2$	CSe	193.4	CSeSe	103.5°
	SeSe	233.5		
$(CF_3)_2S_3$	CS	184.8	SSS	103.8°
	SS	206.5	CSS	103.5°
$(CH_3)_2S_2$	CS	178	CSS	107°
	SS	204		
Cl_2S_2	SCl	207	ClSS	107°
	SS	199		
H_2S_2	SH	133	HSS	95°
	SS	205		
Br_2S_2	SBr	224	SSBr	105°
	SS	198		
Ph_2Se_2	SeC	193	CSeSe	105°, 108°
	SeSe	229		
$(pClC_6H_4)_2Te_2$	TeC	210, 216	CTeTe	93.5°, 95.2°
	TeTe	270.2		
S_3^{2-}		215		103°
S_4^{2-}		203, 207		104.5°
S_8		203.7		107.6°
Se_8		234		105.3°
Te (hexagonal)		282		103.7°
$S_4O_6^{2-}$			SSS	104°, 102°
$S_5O_6^{2-}$			SSS	103°, 106°
SO_2	SO	143.2		119.5°
S_2O	SS	188	SSO	118°
	SO	146		
SeO_2		161	Nonlinear (gas)	
ClO_2		149		118.5°
ClO_2^-		155		110°–111°
$SnCl_2$ (gas)		242		95°*
$SnBr_2$ (gas)		255		95°*
SnI_2 (gas)		273		95°*
$PbCl_2$		246.5		95°*
$PbBr_2$		260		95°*
PbI_2		279		95°*
MgF_2				158°
CaF_2				140°
SrF_2				108°
BaF_2				100°
$SrCl_2$, BaX_2				bent
(X = Cl, Br, I)				

* Assumed.

is unexpected and is not predicted from electron-pair repulsion theory. The tendency to form bent halides follows the trend Ba > Sr > Ca > Mg > Be and F > Cl > Br > I. This is said to indicate the use of d orbitals in bonding, especially for the fluorides, where contraction of the d orbitals is most likely. The Walsh diagram for AB_2 molecules (Fig. 5.7), when modified[5] to include d orbitals, indicates a preference for the bent structure over the linear for the heavier metal atoms.

10.4 THREE COORDINATION

Trigonal Planar

The electron-deficient elements Al, Ga, In, and Tl readily form three-coordinate trigonal planar molecules (Table 10.12) in a similar way to that described for boron (Section 9.7). The few known three-coordinate compounds of divalent mercury are almost trigonal planar, and their existence probably reflects the difficulty for mercury to incorporate all its $6p$ orbitals in four-coordinate bonding, as described previously (p. 172). The ions $HgBr_3^-$ and HgI_3^- have the mercury atoms 30 and 8 pm, respectively, above the plane of the halogens. In the case of $HgBr_3^-$, this is said to result

Table 10.12   Three-Coordinate, Trigonal Planar Compounds

Compound	Bond length (pm)		Bond angle	
$Me_4N^+HgBr_3^-$		252	121.8°, 114.4°, 119.1°	
$(CH_3)_3S^+HgI_3^-$		272, 269, 272	124°, 113°, 123°	
$AlCl_3$ (gas)	Monomeric			
GaI_3		250		
$Ga(CH_3)_3$	Monomeric			
GaF_3 (gas)		188		
$In(CH_3)_3$ (gas)		216		
$[(CH_3)_2Tlpy]^+ClO_4^-$	Probably T-shaped			
SO_3		143		120°
ClF_3	ClF	169.8	FClF'	87.5°
	ClF'*	159.8		
BrF_3 (gas)	BrF	181.0	FBrF'	86.2°
	BrF'	172.1		
BrF_3	BrF	185, 184	FBrF'	82.0°, 88.4°
	BrF'	172		
$PhICl_2$	ICl	245	ClICl	86°
	IC	200		

* F' corresponds to fluorine in the downstroke position of the "T."

from a polar environment around the mercury because a fourth bromine atom is only 290 pm from the mercury whereas the three nearest are at 252 pm.

The trigonal planar shape of sulfur trioxide is indicative of considerable π-electron delocalization, as described for the nitrate ion (p. 146).

The interhalogen compounds also listed in Table 10.12 are planar but T shaped, since there are five electron pairs in the valence shell; two of these pairs are nonbonding. The downstroke of the T is a shorter bond than the other two because it lies in the equatorial plane of the five electron pairs. The XMX' bond angles are all less than 90°, owing to the repulsion effect of the lone pairs.

Pyramidal

All known cases of three-coordinate pyramidal species of the p-block elements have a stereochemically active lone pair, and the central atom is in a low valency state. Trivalent P, As, Sb, and Bi form a variety of pyramidal molecules (Table 10.13), where apparent trends are small and

Table 10.13   The Stereochemistry of Pyramidal Compounds of P, As, Sb, and Bi

Compound	P Bond length (pm)	P Bond angle	As Bond length (pm)	As Bond angle	Sb Bond length (pm)	Sb Bond angle	Bi Bond length (pm)	Bi Bond angle
MH_3	142	93.5°	151.9	91.8°	171	91.3°		
$M(CH_3)_3$	184	98.9°	198	96°				
$M(CF_3)_3$	193	99.6°	205	100.1°	220	100°		
MF_3	154	100°	171	102°	203	81.9° (1)* 104.3° (2)*		
MCl_3	204	100° 6′	216	98.4°	233	99.5°	248	100°
MBr_3	218	101.5°	233	100.5°	251	97°	263	100°
MI_3	243	102°	255	101.5°	267	99°		
$MFCl_2$	155 202	102°						
$M(CN)_3$	178	93° (av)		~90°				

* (1) and (2) correspond to the number of angles.

probably of little significance. Both P and As are effectively six coordinate (in the solid state) in the cyanide compounds $M(CN)_3$ because of three additional long $C-N \cdots M$ interactions (the $N-M$ distance is less than the sum of N and M van der Waals' radii). Elemental P and As occur as the tetrahedral P_4 and As_4 molecules. The bonds from each atom are pyramidal with bond angles close to $60°$, and the P—P distance of 221 pm is intermediate between a single and a double bond. Both these results may arise from either bent σ bonding or involvement of d orbitals. The elemental tetrahedral arrangement persists in the sulfides and oxides of the two elements. For example, the sulfides P_4S_3, P_4S_5, P_4S_7, and P_4S_{10} display progressive loss of P—P bonds but retain the P_4 tetrahedral grouping (Fig. 10.1).

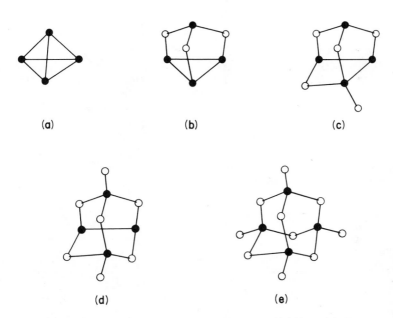

Fig. 10.1   Structures of the phosphorous sulphides: (a) P_4, (b) P_4S_3, (c) P_4S_5, (d) P_4S_7, and (e) P_4S_{10}.

Divalent germanium, tin, and lead also give rise to the pyramidal stereochemistry, in such species as GeF_2 in the solid state, GeF_3^-, $SnCl_2H_2O$, $Sn_2F_5^-$, $SnCl_3^-$, and the $Pb(II)O_3$ groups in Pb_3O_4. Table 10.14 lists a number of other pyramidal molecules. In all cases involving oxygen, some multiple bonding probably occurs. The tetrameric compound $[TlOCH_3]_4$ of thallium(I) has a distorted cubic structure with Tl and O atoms at alternate corners of the cube (Fig. 10.2). In this way the metal achieves three coordination, with the lone pair presumably protruding from the cube corners.

Metal complexes formed with thio- and selenoethers [e.g., $PdCl_2$ $(Et_2S)_2$] involve the donation of a lone pair from the sulfur (or selenium) to the metal atom. This gives three coordination around the S (or Se). The

Table 10.14   Three-Coordinate Pyramidal Compounds

	Compound	Bond lengths (pm)	Bond angles
S	$SOCl_2$	SO 145, SCl 207	ClSCl 114°, OSCl 106°
	$SOBr_2$	SO 145, SBr 227	BrSBr 96°
	SOF_2	SO 141, SF 158.5	FSF 92.8°, OSF 106.8°
	$(CH_3)_2SO$	SO 147, CS 182	CSC 100°, CSO 107°
	$(CH_3)_3S^+I^-$	SC 180–185	CSC 100–104°
	Ph_2SO	SO 147, SC 176	CSC 97.3°, CSO 106°
Se	$SeOCl_2$	SeO 161, SeCl 170?	ClSeCl ~106°, ClSeO ~114°
	$(SeO_2)_n$	SeO_b* 178, SeO_t[†] 173	O_bSeO_b 90°, O_tSeO_t 98°
	SeO_3^{2-}	SeO 176	OSeO 99.3°
Te	TeO_3^{2-}	TeO 188	OTeO 100°
Cl	ClO_3^-	ClO 157, 148	OClO 106.7°
	$AgClO_2$	AgCl 220, ClO 162	OClO 116°
Br	BrO_3^-	BrO 178	OBrO 111.8°
I	IO_3^-	IO 182	OIO 99°
	$HOIO_2$	IO(H) 189, IO 181	OIO 98°
Xe	XeO_3	XeO 176	OXeO 103°

* *b* refers to bridging atom.
† *t* refers to terminal atom.

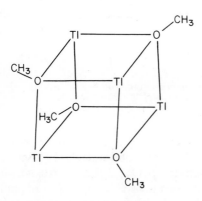

Fig. 10.2   Structure of $(TlOCH_3)_4$.

stereochemistry can be either pyramidal or planar, depending on how the second lone pair on the sulfur (or Se) is used. If it is unused, the stereochemistry will be pyramidal, but if it is involved in a S → metal π bond, the stereo-chemistry will be planar. A recent structure analysis of *trans*-$(Et_2Se)_2PdCl_2$ indicates a pyramidal stereochemistry around the selenium, as does ^1H nmr studies.

10.5   FOUR COORDINATION

Square Planar

Square planar stereochemistry is achieved when the central atom has six valence electron pairs, two of which are nonbonding. Examples occur for the elements Br, I, Xe, and divalent tellurium (Table 10.15). The

Table 10.15   Four-Coordinate, Square Planar Compounds

	Compound	Bond length (pm)	Bond angles
Br	$BrF_4^-(K^+)$	188	
I	$ICl_4^-(K^+)$	242–260	Close to 90°
	I_2Cl_6	ICl_t, 239	84–96°
		ICl_b, 270	
Xe	XeF_4	193	
	$XeF_2 \cdot XeF_4$	Contains square planar XeF_4 groups	
Te(II)	$TeCl_2(thiourea)_2$	TeCl 292, TeS 248 ⎫	cis configuration
	$TeBr_2(thiourea)_2$	TeBr 305, TeS 248 ⎭	
Pb	Pb_2O_3	PbO in PbO_4 groups	
		2 at 194; 1 at 222; 1 at 218	
Hg	$HgCl_2$ 1,4-dioxane	HgO 266, HgCl 234	

Te(II) compounds are interesting; although stereochemically normal, their existence is the result of an increase in the stability of the low oxidation state of the heavier elements.

The I—Cl bond length in ICl_4^- varies with the counter cation, which may be related to the polarizing power of the cation.

Both the $Pb(IV)O_4$ square planar grouping in Pb_2O_3 and the square planar complex $HgCl_2$1,4-dioxane are unexpected. The shape probably results from solid state effects.

Tetrahedral

Tetrahedral compounds fall broadly into three groups: those with four σ-bonding pairs in the valence shell; those having in addition, some π bonding; and those including lone pairs.

The first type, by far the most common, occur mainly for the Group IV elements Si, Ge, Sn, and Pb. A few representative examples are given in Table 10.16. For reasons given earlier (p. 170), four coordinate tetrahedral

Table 10.16 Four-Coordinate Tetrahedral Compounds

	Compound	Bond lengths (pm)	Bond angles
Zn	$ZnCl_4^{2-}$	224	108.3–111.7°
	$ZnCl_2(NH_3)_2$	ZnCl 230, Zn—N 200	
	$ZnBr_4^{2-}$	238–241	106.4–114.9°
	$ZnCl_2(thiourea)_2$	ZnCl 232, ZnS 235	SZnS 111.5°
			ClZnCl 107.3°
	$ZnCl_2$	227–234	Tetrahedral
Cd	$Cd(NH_3)_4^{2+}$	210	
	$CdCl_2(thiourea)_2$	CdCl 251, CdS 245	ClCdCl 103°
			ClCdS 106°, SCdS 129°
Hg	$Hg(SCN)_4^{2-}$	255	109°
	HgI_4^{2-}	268–280	105–111°
	$HgCl_2(Ph_3AsO)_2$	HgCl 232, 233	ClHgCl 146.5°
		HgO 232, 237	OHgO 92.5°
Al	$AlCl_4^-$	213	Tetrahedral
	Al_2Br_6	$AlBr_t$* 223–233	Br_bAlBr_b 98°
		$AlBr_b^†$ 234, 242	All others, 109–115°
	$AlBr_3H_2S$	AlBr 234, AlS 240	
	$(CH_3)_4Al_2Cl_2$	AlC 193, $AlCl_b$ 231	ClAlCl 89°, CAlC 127.5°
Ga	$GaCl_4^-$ in "$GaCl_2$"	219	Tetrahedral
	Ga_2X_6	GaCl 209, 229	Distorted tetrahedron
	$(X = Cl, Br)$	GaBr 225, 235	
In	$InCl_4^-$		Tetrahedral
	In_2X_6	InCl 246	
	$(X = Cl, Br, I)$	InBr 258 (av)	
		InI_b 284	I_bInI_b 93.7°
		InI_t 264	I_tInI_t 125.1°
Tl	$TlBr_4^-$	251	Tetrahedral
	TlS	Tl(III)S 260	Distorted tetrahedron
Si	$SiCl_4$	201	Tetrahedron
	$SiClF_3$	SiCl 203, SiF 155	FSiF 108.5°
	Si_2Cl_6O	SiO 164, SiCl 202	ClSiCl 109.5°
	$(CH_3)_2SiBr_2$	SiC 192, SiBr 221	BrSiBr 109.5°
Ge	$GeBr_4$	229	
	$GeHCl_3$	GeH 155, GeCl 211	ClGeCl 108.3°
	CH_3GeH_3	GeH 153, GeC 195	HGeH 109° 15′
Sn	$SnBr_4$	237–244	Tetrahedron
	CH_3SnI_3	SnC 217, SnI 268	ISnI 109.5°
	$(CH_3)_3SnCl$	SnC 219, SnCl 237	CSnC 111°
Pb	$Pb(CH_3)_4$	220	Tetrahedron
	$Pb_2(CH_3)_6$	PbPb 288, PbC 225	Tetrahedron
P	PCl_4^+	198	Tetrahedron
	PH_4^+	142	Tetrahedron
	$H_3BP(NH_2)_3$	PN 165, PB 189	BPN 123.3°, 109°
			Rest 99.6–116.8°
	$PBr_4^+(Br_3^-)$	217	BrPBr 107.3–110.4°
As	$(CH_3)_4As^+$		Tetrahedron
	Ph_4As^+	195	Tetrahedron

* t = Terminal.   † b = Bridge.

Table 10.17 Attainment of Four σ Pairs

Elements	Valence state configuration	Number of ligands in addition to those required to satisfy normal valency
Zn, Cd, Hg(II)	s^1p^1	2, e.g., Cl^-, Ph_3AsO
Al, Ga, In, Tl(III)	$s^1p^1p^1$	1, e.g., Cl^-, dimerization, electron donor, e.g., NH_3
Si, Ge, Sn, Pb(IV)	$s^1p^1p^1p^1$	None
Sn(II)	$s^2p^1p^1$	2. An electron donor, e.g., Cl^-, and an electron acceptor, e.g., transition metal
P, As, Sb(III)	$s^2p^1p^1p^1$	1, Electron acceptor, e.g., Cl^+, H^+, transition metal

Table 10.18 Four-Coordinate Tetrahedral Compounds

	Compound	Bond length (pm)	Bond angle
Si, Ge	Silicates $[(CH_3)_2Si]_2O$ $[GeO_3^{2-}]_n$		Tetrahedral
P	POF_3	PF 152, PO 145	FPF 102.5–106°
	$POCl_3$	PCl 199, PO 145	ClPCl 103.5°
	$PSBr_3$	PBr 213, PS 189	BrPBr 106°
	P in P_4O_{10}, P_4O_6		OPO 100–116°
	$[HPO_3]^{2-}$	PO 151	OPO 110°
	$(PNCl_2)_3$	PN 165, PCl 197	ClPCl 108.5°
	$(C_2H_5)_3PS$	PS 186, PC 187	SPC 112°, CPC 107°
As	$(AsO_3)^{n-}$	160	Tetrahedral
	$(CH_3)_2AsO(OH)$	AsC 119, AsO 162	106–115°
S	SO_2Cl_2	SO 143, SCl 199	OSO 119.8°, ClSCl 111.2°
	SO_4^{2-}	SO 142, 155	98–117°
	$(CH_3)_2SO_2$	CS 180, SO 143	CSC 115°, CSO 105°
	$S_2O_3^{2-}$	SS 196–201	
Se	SeO_4^{2-}	SeO 161	Tetrahedral
Cl	ClO_4^-	150	Tetrahedral
	ClO_3F		Distorted tetrahedron
	$HOClO_3$	ClO(H) 164, ClO 142	HOClO 100°
	Cl_2O_7	ClO_t* 142, ClO_b† 172	97–115°
I	IO_4^-	179	Tetrahedral

* t = Terminal.
† b = Bridge.

compounds of Zn, Cd, and Hg are most prolific for zinc. The tetrahedral stereochemistry is achieved for MX_3 compounds through dimerization. The shape is distorted by elongation along the molecular axis, probably because of repulsion between the metal atoms. For Group V, tetrahedral compounds of the first type are few and are restricted to phosphorus and arsenic (Table 10.16). However, the compounds $R_3M(M = P, As, and Sb)$ coordinate through M to form metal complexes. Compare the wide use of Ph_3P as a ligand in transition metal chemistry. The Group V elements therefore achieve a distorted tetrahedral environment. The same situation applies to the three-coordinate anion SnX_3^- ($X = Cl, Br$) when used as a ligand, as in $[PtCl_2(SnCl_3)_2]^{2-}$ and $[Rh(SnCl_3)_4]^{3-}$. The methods of achieving four σ pairs for the various elements are summarized in Table 10.17.

A number of compounds (Table 10.18) are known which are tetrahedral and have some multiple bonding. They occur for elements of Groups V, VI, and VII having more than four valence electrons. The extra electrons are used in multiple bonding, particularly to oxygen, sulfur, and occasionally to nitrogen. For example, in $POCl_3$ the π component of the P—O bond will be $P_{d\pi}$-$O_{p\pi}$ overlap.

Distorted tetrahedral compounds are formed by the p-block elements in low valency states. In all examples (Table 10.19), the lone pair occupies a fifth position around the central atom, normally one of the equatorial positions of a trigonal bipyramidal shape. The gallium compound [Ga(di-

Table 10.19 Four-Coordinate Distorted Tetrahedral Compounds

	Compound	Bond lengths (pm)		Bond angles	
Ga(I)	in				
	$[Ga(dioxime)_2]^+Cl^-$				
Sn(II)	SnO		221	OSnO	75°
Pb(II)	PbO		230		
As(III)	$AsCl_4^-$				
Sb(III)	$Sb_2F_7^-$				
	(Cs^+)				
	$Sb(OH)_4^-$			OSbO	165°
S(IV)	SF_4			Distorted tetrahedron	
Se(IV)	SeF_4		177	FSeF	120°, 104.5°
	Ph_2SeCl_2	SeCl	230	ClSeCl	180°
	Ph_2SeBr_2	SeBr	252, SeC 191	BrSeBr 180°, CSeC 110°	
Te(IV)	$TeCl_4$		233	ClTeCl	186°, 90–120°
	$(CH_3)_2TeCl_2$	TeCl	248, 254	ClTeCl	172.3°
		TeC	208, 210	CTeC	98.2°
I	IF_4^+ in $SbF_6^- IF_4^+$			Distorted tetrahedron	

oxime)$_2$]$^+$Cl$^-$ is unusual in that the lone pair appears to occupy an axial position of a trigonal bipyramid.

The structure of the oxides SnO and PbO appear to be based on a square pyramidal shape; this may be due to greater ionic character of the bonds leading to a layer lattice structure. A weak Sn—Sn interaction (370 pm) is also present in SnO.

10.6   FIVE COORDINATION[6]

Trigonal Bipyramid

As discussed earlier (Section 4.5) the two axial bonds in a trigonal bipyramidal compound MX_5 are longer than the equatorial ones. Generally, the most electronegative ligands preferentially occupy the axial positions in mixed ligand compounds; for example, from nmr studies it is clear that the compounds PF_3Cl_2, PF_4R_2N, PF_3R_2, PF_2R_3, and PHF_3R all have fluorine atoms in the axial positions.

A number of compounds that are five coordinate in the gaseous state associate in the liquid or solid state. For example, both SbF_5 and SbF_3Cl_2 are partially associated in the liquid state. Therefore, be cautious when inferring the coordination number and stereochemistry of compounds in condensed states from a knowledge of their structures in the gaseous state.

Nuclear magnetic resonance studies of the fluorides of phosphorus have revealed a marked degree of structural mobility. The molecular structure of PF_5 is typical of a trigonal bipyramidal compound, but nmr studies indicate that all the fluorines are in equivalent environments, suggesting they readily exchange positions (faster than the nmr time scale; see Table 1.5 and Section 1.4). A mechanism for the interconversion of fluorine positions has been proposed by making use of the normal modes of vibration of a molecule with the D_{3h} point group (Fig. 10.3).

It is estimated that an activation energy of only 24 to 28 kJ mol^{-1} is required for such a rearrangement. The transition state (Fig. 10.3) corresponds

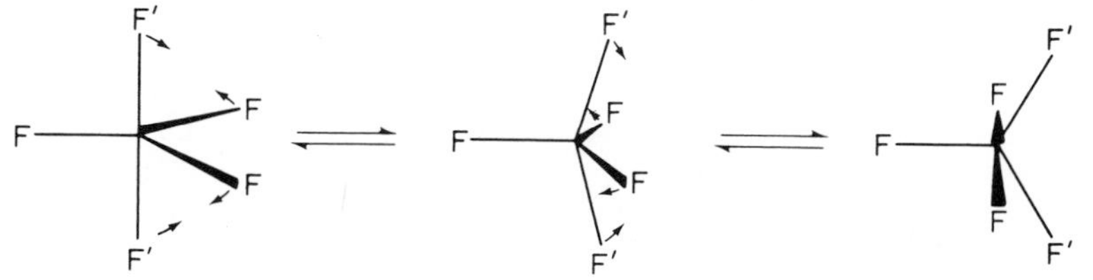

Fig. 10.3   Vibrational interconversion of the fluorines in a trigonal bipyramidal structure.

to a pyramid and indicates the close relationship, in stereochemistry and energy, of the two normal or limiting stereochemistries for five coordination. Many compounds have structures intermediate between the limiting structures. In such cases, it is of little value to discuss which ideal structure better fits the experimental facts, inasmuch as the actual shape taken up by a species is often determined by such factors as crystal packing and steric requirements of the ligands, rather than by any subtle advantages in bonding. For example, the tridentate ligand $\alpha\alpha'\alpha''$ terpyridyl bonds as a planar ligand and can only span from 150 to 160° between the terminal nitrogens, which bond in the axial positions, as in $ZnCl_2$(terpy) (Fig. 10.4).

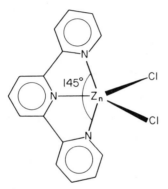

Fig. 10.4   Structure of $ZnCl_2$ (terpyridyl).

Zinc forms the greatest number of five-coordinate compounds of the triad Zn, Cd, and Hg, and the stereochemistry is generally trigonal bipyramidal. Since an outer d orbital is used for bonding, it is to be expected that zinc will form five-coordinate compounds more readily than mercury. It is difficult, however, to understand the absence of well-characterized five-coordinate compounds of cadmium, but this may be due to lack of study.

Certain zinc compounds actually dimerize (Table 10.20) to give distorted trigonal bipyramidal structures, rather than remain monomeric with the more usual tetrahedral structure.

Table 10.20   Five-Coordinate Trigonal Bipyramidal Stereochemistry

Compound	Bond lengths (pm)	Bond angles	
Zn $ZnCl_2$(terpy)	ZnCl 229, ZnN 211–234	NZnN	151°
(Zn(diethyldithio-	ZnS_{ax} 244, 281		
carbamate)$_2$)$_2$	ZnS_{eq} 235		
[Zn(bisNmethylsalicyl-	ZnO_{ax} 211	$O_{ax}ZnO_{ax}$	177°
aldimine)$_2$]$_2$	ZnO_{eq} 202, ZnN_{eq} 201	NZnN	123°
NaZn(OH)$_3$	ZnO_{ax} 265, ZnO_{eq} 198		

Table 10.20 (*cont.*)

Compound	Bond lengths (pm)	Bond angles	
Al	AlH$_3$(TMED)	AlN 219, 234	NAlN	176–178°
	AlH$_3$((CH$_3$)$_3$N)$_2$	AlN 218		
In	In(CH$_3$)$_3$	InC$_{eq}$ 212	C$_{eq}$InC$_{eq}$	117–122°
		InC$_{ax}$ 311, 359	C$_{ax}$InC$_{ax}$	163°
Sn	[SnCl$_5$]$^-$	SnCl$_{eq}$ 238		
		SnCl$_{ax}$ 237, 239		
	[(CH$_3$)$_3$SnF]$_n$	SnF$_{ax}$ 210–260		
		SnC$_{eq}$ 210		
	(CH$_3$)$_3$SnCl pyridine	SnCl 242		
P	PF$_5$ (gas)	PF$_{eq}$ 153.4		
		PF$_{ax}$ 157.7		
	CH$_3$PF$_4$ (gas)	PF$_{eq}$ 154.3	F$_{ax}$PC	91.8°
		PF$_{ax}$ 161.2	F$_{eq}$PC	122.2°
		PC 178		
	(CH$_3$)$_2$PF$_3$ (gas)	PF$_{eq}$ 155.3	F$_{ax}$PF$_{eq}$	88.9°
		PF$_{ax}$ 164.3, PC 180	F$_{eq}$PC	118°
	PCl$_5$ (gas)	PCl$_{eq}$ 204		
		PCl$_{ax}$ 219		
	PCl$_2$F$_3$	PF$_{eq}$ 159		
		PF$_{ax}$ 205		
	Ph$_5$P	PC$_{eq}$ 185		
		PC$_{ax}$ 199		
As	Ph$_5$As	Isomorphous with Ph$_5$P		
	AsF$_5$	Trigonal bipyramid	(IR, Raman)	

			BrAsBr	178.4°
Sb	SbCl$_5$ (vapor)	SbCl$_{eq}$ 231		
		SbCl$_{ax}$ 243		
	SbCl$_5$ (solid)	SbCl$_{eq}$ 229		
		SbCl$_{ax}$ 234		
	(CH$_3$)$_3$SbX$_2$			
	(X = Cl, Br, I)	SbCl$_{ax}$ 249 ⎫		
		SbBr$_{ax}$ 263 ⎬SbC 210		
		SbI$_{ax}$ 288 ⎭		
	(ClCH=CH)$_3$SbCl$_2$	SbCl$_{ax}$ 245, SbC 215		
	Ph$_2$SbCl$_3$	SbCl$_{ax}$ 252		
		SbCl$_{eq}$ 243		
	Ph$_3$SbCl$_2$	SbCl$_{ax}$ 250		
Bi	Ph$_3$BiCl$_2$	Trigonal bipyramid		
S	SOF$_4$ ⎫	Trigonal bipyramids	IR	
	SF$_4$(NCF$_3$) ⎭		NMR	

The trihydride compounds of aluminum, $AlH_3(N(CH_3)_3)_2$ and $AlH_3(TMED)$ do not have bridging H atoms, and the aluminum is in a trigonal bipyramidal stereochemistry. Trimethyl indium(III) is trigonal planar (In—C, 212 pm), but two further methyl groups occur in the axial positions (In—C, 311 and 359 pm). In such cases, it is not always easy to decide whether the long In—C distances are indicative of bonding.

A number of five-coordinate tin(IV) compounds (Table 10.20), such as $[SnX_5]^-$, $[SnX_3Y_2]^-$, SnX_4L, and SnX_3Z (where Z acts as a bridging group in the axial positions), have the trigonal bipyramidal structure. Stoichiometry is not always a guide to coordination number; for example, the compound $Me_4Sn_2Cl_4(terpyridyl)$ contains both a six-coordinate cation $[Me_2SnCl(terpyridyl)]^+$ and a five-coordinate trigonal bipyramidal anion $[Me_2SnCl_3]^-$.

The Group V elements P, As, Sb, and Bi form a large number of pentavalent five-coordinate compounds, all trigonal bipyramidal except for Ph_5Sb. In a number of cases, the compounds have been studied in the gaseous state, or in nonsolvating solvents. In the condensed states, some association often takes place. In the case of PX_5 (X = Cl, Br), salts are formed in the solid state, viz., $PX_4^+PX_6^-$. For the series of compounds PF_5, CH_3PF_4, and $(CH_3)_2PF_3$, the difference between the P—F equatorial and axial bond lengths increases with the number of methyl groups (Table 10.21). This may

Table 10.21   Bond Lengths PF (pm) in $(CH_3)_nPF_{5-n}$ Compounds

Compound	Equatorial	Axial	Difference
PF_5	153.4	157.7	4.3
CH_3PF_4	154.3	161.2	6.9
$(CH_3)_2PF_3$	155.3	164.3	9.0

result from the lower electronegativity of the CH_3 group enabling the fluorine to have a greater share of the bonding electrons. Increased electron density in the PF bonds will lead to greater bond-pair repulsions, causing a lengthening of the P—F bonds and particularly the axial ones.

Five coordination for the elements of Group VI [e.g., S in SOF_4 and $SF_4(NCF_3)$] is achieved by the formation of multiple bonding ($S{=}O$ and $S{=}N$, respectively). The double bonds occur in the equatorial positions.

Square Based Pyramid

The square pyramidal stereochemistry can arise from two causes: distortion from the trigonal bipyramid, Table 10.22, or presence of a lone pair on the central atom in addition to the five σ bonding pairs (Table 10.23).

Table 10.22 Five-Coordinate Square Pyramidal Stereochemistry

	Compound	Bond lengths	Bond angles
Zn	Zn(acac)$_2$H$_2$O	ZnO 202	Distorted square pyramid
		Cd compound isomorphous	
	Zn(glutonate)(H$_2$O)$_2$	ZnO 203, ZnN 210	Regular square pyramid
	Zn(Ph$_2$MeAsO)$_4$(ClO$_4$)$_2$	Isostructural with Co compound	
	Zn(NN disalicylidenepropylene-1,2-diamine)H$_2$O	ZnO$_{base}$ 194 ZnOH$_{2ax}$ 213 ZnN$_{base}$ 208	Distorted square pyramid
Hg	Hg$_2$O$_2$NaI	HgI 333, HgO 203	
	Hg(SMe)$_2$	HgS 236, HgS 326	Polymer, distorted square pyramid
As	PhAs(diethyldithiocarbamate)$_2$	AsC 195 AsS 233, 287	SAsS 84.6°, 68° CAsS 138°

Table 10.23 Five-Coordinate Square Pyramidal Stereochemistry

	Compounds	Bond lengths (pm)*	Bond angles	
In	InBr	InBr 329, InBr$_{ap}$ 280		
	InI	Similar		
	[InCl$_5$]$^{2-}$	Square pyramid		
Tl	TlI	TlI$_{base}$ 349, TlI$_{ap}$ 336		
Sb	(NH$_4$)$_2$SbCl$_5$	SbCl$_{base}$ 262, SbCl$_{ap}$ 236		80–90°
	K$_2$SbF$_5$	SbF$_{base}$ 202, SbF$_{ap}$ 208		82–87°
	SbPh$_5$	SbC 205–235	C$_{base}$SbC$_{base}$	88°
		(no long apical bond)	C$_{base}$SbC$_{ap}$	102°
	KSbF$_4$, i.e.,			
	(SbF$_4^-$)$_4$	SbF$_{base}$ 224, SbF$_{ap}$ 201		
	SbCl$_3$(Ph$_3$AsO)$_2$	SbCl$_{base}$ 259, SbCl$_{ap}$ 220 SbO$_{base}$ 226, 227	All close to 90°	
Bi	[BiCl$_5$]$^{2-}$	BiCl 261–287		
	[Bi$_2$Cl$_8$]$^{2-}$	Two BiCl$_5$ groups		
	In BiCl$_3$/Bi melt	Share base edge, BiCl 261–278		
Se	SeOCl$_2$(py)$_2$	SeO$_{ap}$ 159, SeCl 257, 239	ClSeCl	165.2°
		SeN 219	NSeN	173.8°
Cl	ClF$_5$	ClF$_{base}$ 172, ClF$_{ap}$ 162		
Br	BrF$_5$	BrF$_{base}$ 178, BrF$_{ap}$ 168		80–87°
I	IF$_5$	IF$_{base}$ 186, IF$_{ap}$ 175		
Xe	XeOF$_4$	Square pyramid		

* *ap* = Apical.

The latter type occurs for the *p*-block elements in low oxidation states and particularly for the heavier elements (especially antimony), where the low oxidation state is relatively more stable.

For the majority of square pyramidal structures studied, which have a lone pair, the apical bond is shorter than the bonds in the square base. Prediction of bond lengths is not as straightforward as for a trigonal bipyramidal structure, and factors such as lattice interactions and ligand influence affect the result.

The two species Ph_5Sb and $[InCl_5]^{2-}$ do not have a lone pair but are square pyramidal. Distortion from a trigonal bipyramidal structure is not expected, and the reason for the shape is not at all clear.

10.7   SIX COORDINATION

Octahedral Stereochemistry

Six-coordinate compounds are almost invariably octahedral for the *p*-block elements. Distortions may be expected when the number of σ-valence electron pairs is greater than six. Six coordination is very common, and examples can be found for most elements considered in this chapter (Tables 10.24 and 10.25 list a few representative examples).

Table 10.24   Six-Coordinate Octahedral Compounds

Compound	Bond lengths (pm)	Bond angles
$Zn(acetate)_2(H_2O)_2$	ZnO_{ac} 217, 218	Distorted octahedron
	ZnO_{H_2O} 214	
$Zn(biuret)_2Cl_2$	ZnCl 253	Close to 90°
	ZnO 205, 203	
$Zn(N_2H_4)_2Cl_2$	ZnN 215, ZnCl 258	N_2H_4 bridges
$[CdCl_6]^{4-}$	CdCl 253, or 264	
$CdCl_2(urea)_2$	CdCl 264, CdO 228	Cl bridges, angles $\sim 90°$
$CsCdCl_3$ $\Big\}$	CdCl 264	Share faces
Cs_2CdCl_4		Share corners
$KHg(CN)_2I$	HgC 208, HgI 338	CN-Hg-CN linear
$Al(BH_4)_3$	AlH 210, Al—B 215	Distorted octahedron
AlF_6^{3-}	AlF 181	
$AlCl_3$	$CrCl_3$ structure	
Al_2O_3	Close-packed O^{2-}, Al^{3+} in	
	octahedral holes	
GaF_3	GaF 189	
Cs_3GaF_6		Octahedral
Ga_2O_3	α and γ form, Ga^{3+} in octahedral	
	holes	

Table 10.24 (*cont.*)

Compound	Bond lengths (pm)	Bond angles
$(NH_4)_2[InCl_5H_2O]$	InO 223, InCl 258	
$InPO_42H_2O$	4InO 212, 2InO 225	H_2OInH_2O 134° waters cis
$[TlCl_6]^{3-}$	TlCl 249	
$[TlBr_6]^{3-}$	TlBr 259	
$Cs_3[Tl_2Cl_9]$	TlCl$_t$ 254	Cl$_t$TlCl$_t$ 96°
	TlCl$_b$ 266	Cl$_b$TlCl$_b$ 79°
SiF_6^{2-}	SiF 169.5	
GeF_6^{2-}	GeF 172–177; varies with cation	
GeO_2	4GeO 187, 2GeO 191	TiO_2 structure
SnO_2	4SnO 205, 2SnO 206	Rutile structure
$SnCl_6^{2-}$	SnCl 239–245	
SnI_6^{2-}	SnI 284	Antifluorite structure
SnF_6^{2-}	SnF 183(2), 192(2), 196(2)*	Distorted octahedron
Me_2SnF_2	SnC 208, SnF 212	
$SnCl_42SeOCl_2$	SnCl 241, 236, SnO 212	81–100°
$SnCl_42Me_2SO$	SnCl 236–247, SnO 217, 210	82–100°
$PbCl_6^{2-}$	PbCl 250	
Pb_3O_4	$Pb^{IV}O_6$ octahedra	
	PbO 215	
PX_6^- $(X = F)$	PF 173, (H$^+$); 158 (2), 173(4) (Na$^+$)	
$(X = Cl)$	PCl 207	
AsF_6^-	AsF 180, (K$^+$) 188 (Li$^+$)	Varies with cation
$SbCl_5POCl_3$	SbCl 233, SbO 218	
$SbCl_5DMF^†$	SbCl 234, SbN 205	Close to 90°
SF_6	SF 156–158	
$S_2F_{10}O_2$	SF 156, SO 166	
SF_5OF	SF 153, SO 164	90°
SeF_6	SeF 167 or 170	
TeF_6	TeF 182 or 184	
$[TeO_6]^{6-}$	TeO 196	
$[IO_6]^{5-}$	IO 185	
$[XeO_6]^{4-}$	XeO 184–186	

* Numbers in parentheses refer to number of bonds.
† DMF = dimethylformamide.

Aluminum achieves six coordination with the small ligand fluorine in the species AlF_6^{3-}, $(AlF_5^{2-})_n$, $(AlF_4^-)_n$, and crystalline $(AlF_3)_n$. The dimeric anion $[Tl_2Cl_9]^{3-}$ is an example of a number of ions $[M_2(III)X_9]^{3-}$ in which the metal atom is six coordinate with a distorted octahedral environment (Fig. 10.5). The degree of distortion varies with the central metal ion; for some transition metals, direct metal-metal bonds are formed.

The distortion in the anion SnF_6^{2-} is unexpected and may arise from crystal lattice interactions. Cations can influence the $A—X$ bond lengths in

Table 10.25   Six-Coordinate Octahedral Compounds

Compound	Bond lengths (pm)	Bond angles
TlF	2TlF 259, 2TlF 275, 2TlF 304	Distorted octahedron
	TlCl and TlBr similar	
GeI_2	CdI_2 structure	
GeS	GeS 247(1), 264(2), 291(1), 300(2)*	Distorted octahedron
SnS	SnS 262(1), 268(2), 327(2), 339(2)	Distorted octahedron
PbI_2	CdI_2 structure	
Pb_2O_3	$Pb^{II}O_6$ octahedron	
	PbO 224(2), 254(2), 281(2)	Distorted octahedron
AsI_3	AsI 3 short, 3 long	
$Cs_3As_2Cl_9$	AsCl 225	ClAsCl 96°
As(NNdiethyldithio-carbamate)$_3$	AsS 235(3), 280–290(3)	Distorted trigonal prism
SbI_3	SbI 3 short, 3 long	
$[SbX_6]^{2-}$,	Contains SbX_6^{3-} and SbX_6^-	Undistorted octahedra
$(X = \text{halogen})$		
$Cs_3Sb_2Cl_9$	Isostructural with As compound	
BiF_3	αUF_5 structure	
$[BiCl_6]^{3-}$	BiCl 252	
$[SeX_6]^{2-}$,	SeCl 238–241; varies with cation ⎫	Regular octahedra
$(X = \text{Cl, Br})$	SeBr 254 ⎭	
$(SeOCl_2)_2SnCl_4$	Se 6 coordinate	Distorted octahedron
	SeO 173, SeCl 300–340	
$[TeX_6]^{2-}$,	TeCl 248–251; varies with cation ⎫	Regular octahedron
$(X = \text{Cl, Br})$	TeBr 262 ⎭	
XeF_6		Distorted octahedron

* Numbers in parentheses refer to number of bonds.

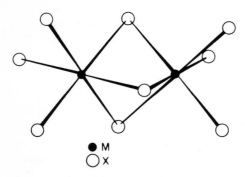

● M
○ X

Fig. 10.5   Structure of the $[M_2X_9]^{3-}$ species.

halogeno-anions such as AX_6^{n-}. It appears that small strongly polarizing cations can lengthen bonds, presumably owing to a strong electrostatic attraction between the ligands of the anion and the cation (Table 10.26).

Table 10.26 Influence of Cation on Bond Lengths

Compound	As—F	Compound	Sb—F
$KAsF_6$	180	$NaSbF_6$	178
$LiAsF_6$	188	$LiSbF_6$	189

Except for the hexahalogeno-anions of Se and Te, there is insufficient precise information available to discuss the stereochemical influence of the lone pair in the six-coordinate compounds listed in Table 10.25. The majority of compounds show some distortion, although this may be the result of polymerization in the case of compounds such as the thallium(I) halides and crystalline trihalides of Group V.

The shapes expected for seven electron pairs, one of which is a lone pair, are given on p. 68. However, for SeX_6^{2-} and TeX_6^{2-} ($X = $ Cl, Br), the stereochemistry is a regular octahedron, indicating that the lone pair is not stereochemically active (p. 68). More structural work is required on similar compounds before a definitive answer can be given to the role of the lone pairs in six-coordinate compounds. For example, a knowledge of the structure of IF_6^- would be interesting.

10.8 HIGH COORDINATION NUMBERS

There are a few definitive examples of the high coordination numbers 7, 8, and 9 for the *p*-block elements. The stereochemistry expected for these coordination numbers has been discussed in Chapter 4.

Seven Coordination

A few examples of seven-coordinate structures are known (Table 10.27). The zinc and cadmium compounds corresponding to $PbCl_2$ (thiourea)$_2$ are four coordinate. The heptafluoride IF_7 is probably a pentagonal bipyramid, although there has been controversy over the shape at various times.

Eight Coordination

The dominant stereochemistry for eight coordination is the dodecahedron. Perhaps eight-coordinate compounds, as well as seven- and nine-coordinate compounds, are best considered as ionic rather than

Table 10.27   Seven-, Eight-, and Nine-Coordinate Compounds

Compound	Bond lengths	Bond angles
	Seven Coordination	
$PbCl_2(thiourea)_2$	PbCl 328, 317, 275	Distorted trigonal prism
	PbS 302, 304, 310, 292	Angles 62–95°
$Pb(acetate)_2(thiourea)$	PbS 310, 310, 334	Pentagonal pyramid
	PbO 263–273(4)	
IF_7	IF_{eq} 183, IF_{ax} 194	Pentagonal pyramid
	Eight Coordination	
$K_3Hg(NO_2)_4NO_3$	HgO 240	
$Cd(NO_3)_24H_2O$	CdO_{H_2O} 226, 233	Distorted
	CdO_{NO_3} 244, 259	Dodecahedron
Ga^+ in $Ga^+[GaCl_4]^-$	GaCl 320–330	
Tl^+ in $Tl^+[TlBr_4]^-$	TlBr 346 (long)	Irregular dodecahedron
$TlNO_3(thiourea)_4$	TlS 343 (long)	Cube?
PbF_2 (cryst)	PbF 257	CaF_2 structure
$Pb(acetate)_4$		Dodecahedron
	Nine Coordination	
$SnCl_2$ (cryst)	SnCl 322, 266, 330, 278(2), 306(2), 386(2)	Tripyramid
PbX_2		
(X = F, Cl, Br)		Tripyramids
Pb(OH)Cl	PbCl 323(5)	
	PbO(4) 267, 293	
PbFCl		Tripyramid

covalent. The ionic eight-coordinate radii of In(III), Sn(IV), and Pb(IV) lie in the range of from 150 to 165 pm, which is large enough to give a radius ratio for eight-coordination with anions having radius less than 200 pm.

Nine Coordination

The stereochemistry observed for nine coordination is the tripyramid (Fig. 4.8); it occurs for tin and lead in their divalent oxidation states. Interestingly, the high coordination numbers 7, 8, and 9 occur for a number of p-block elements in their low oxidation states [e.g., Tl(I). Ga(I), Sn(II), and Pb(II)]. This suggests that the lone pair is not limiting the co-ordination number displayed by the elements.

BIBLIOGRAPHY

1. F. A. COTTON and G. WILKINSON, *Advanced Inorganic Chemistry*, 3rd ed., Interscience Publishers, New York, 1972.

2. J. D. DUNITZ and L. E. ORGEL, *Advan. Inorg. Chem. Radiochem.*, **2**, 1 (1960).

3. L. E. ORGEL, *J. Chem. Soc.*, 4186 (1958).

4. Structural parameters in Table 10.9 and following tables are obtained from L. E. SUTTON, ed., *Interatomic Distances, Chem. Soc. London Spec. Publ.* 11 (1958), *Suppl. Spec. Publ.* 18 (1965); and A. F. WELLS, *Structural Inorganic Chemistry*, 3rd ed., Clarendon Press, Oxford, Engl., 1962. More recent data are obtained from the literature.

5. E. F. HAYES, *J. Phys. Chem.*, **70**, 3470 (1966).

6. E. L. MUETTERTIES and R. A. SCHUNN, *Quart. Rev. London*, **20**, 245 (1966).

FURTHER READING

W. E. DASENT, *Nonexistent Compounds*, Arnold, Ltd., London, 1965.

C. S. G. PHILLIPS and R. J. P. WILLIAMS, *Inorganic Chemistry*, Vols. 1 and 2, Clarendon Press, Oxford, Engl., 1965.

11

The Transition Metals

Ti	V	Cr	Mn	Fe	Co	Ni	Cu
Zr	Nb	Mo	Tc	Ru	Rh	Pd	Ag
Hf	Ta	W	Re	Os	Ir	Pt	Au

We shall discuss the stereochemistry of the transition metals according to their coordination numbers, except that metal-metal cluster compounds and π complexes will be considered independently. Previous statements concerning the factors that determine the coordination number of a central atom apply equally well to the transition metals (cf. Sections 8.2, 9.2, and 10.1).

A major difference in stereochemistry between the nontransition and transition elements is that the lone-pair effect for the regular elements is not operative in the same way for the transition metals. For example, the complex cation $[Co(NH_3)_6]^{3+}$ has nine electron pairs in the valence shell, as follows:

$$\underbrace{3d^6}_{\substack{\text{metal } d \\ \text{electrons}}} \quad \underbrace{3d^4 \; 4s^2 \; 4p^6}_{\substack{\text{from six} \\ \text{ligands}}}$$

three of which are nonbonding, yet the ion is a regular octahedron. The reason for the lack of distortion is that the d orbitals are not fivefold degenerate in the field of the ligands and the six metal d electrons reside in the d_{xy}, d_{yz}, and d_{zx} orbitals, which together have spherical symmetry with respect to the six ligands. However, if the nonbonding electrons do not have spherical symmetry, the compound may have a distorted stereochemistry (Section 7.4), but the distortion is not of the same magnitude or type as arising from the lone pairs actually occupying a stereochemical position around the central atom.

11.1 TWO COORDINATION

Two coordination is rare for the transition metals in compounds in the solid state; it is restricted to the chemistry of Cu(I), Ag(I), Au(I), which all have the d^{10} configuration. The chemistry is therefore similar to that of Zn(II), Cd(II), and Hg(II) (p. 170). In all cases listed in Table 11.1, the compounds

Table 11.1 Linear Compounds

Cu*	Ag	Au*
Cu$_2$O†	Ag$_2$O†	AuCN†
	AgCN†	[AuCl$_2$]$^-$
	AgSCN†	[Au(CN)$_2$]$^-$
	[AgCl$_2$]$^-$	AuI†
	[Ag(CN)$_2$]$^-$	[AuClPCl$_3$]
	[Ag(NH$_3$)$_2$]$^+$	
	[Ag(C$_3$F$_7$CO$_2$)]$_2$	

* Other compounds are probably linear but have not been confirmed structurally.
† Through polymerization.

are linear, or nearly so. Polymeric Ag(SCN) is an exception; the SAgN bond angle is 165°. The angular structure arises from the different bonding modes of the two ends of the thiocyanate ligand. The linear Ag—N≡CS system is imposed by the bonding requirements of the nitrogen, whereas the Ag—S—CN system is bent, owing to the influence of the lone pairs on the sulfur atom.

Two coordination is achieved in one of two ways: polymerization of simple MX compounds, where X is a monovalent anion; coordination of a second ligand X or L (neutral) to MX, to give monomeric species $[MX_2]^-$ or MXL or $[ML_2]^+X^-$.

The predominance of linear two-coordinate compounds of gold has been related to the ns-np energy separation, which is 3.79 eV (Cu), 3.66 eV (Ag), and 4.63 eV (Au). It is more difficult, therefore, for gold to attain three or four coordination, which requires the incorporation of more p orbitals of

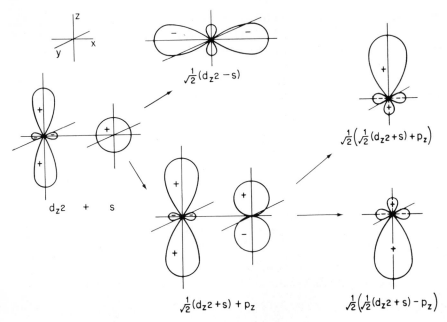

Fig. 11.1   Hybrids composed of s, p_z, and d_{z^2} orbitals.

high energy. An alternative explanation has been proposed, in which the metal ion uses hybrids composed of s, d_{z^2}, and p_z orbitals, as in Fig. 11.1. The two d electrons in the metal d_{z^2} orbital can be placed in the $1/\sqrt{2}(d_{z^2} - s)$ hybrid, whereas the other hybrid $1/\sqrt{2}(d_{z^2} + s)$ can combine with the p_z orbital to give two new linear hybrids. The $d^{10} - d^9 s$ energy separation is therefore important. It is 2.7 eV, 4.8 eV, and 1.9 eV for Cu(I), Ag(I), and Au(I), respectively. This approach suggests that Cu(I) would form linear two-coordinate compounds more readily than Ag(I), yet the reverse seems to be the case. Finally, it should be pointed out that neither explanation can suggest why Ag(I) is good at forming linear compounds.

The dihalides of the transition metals in the vapor state are also likely to be two coordinate and linear. For example, gaseous $CrCl_2$ is mainly monomeric and linear, and its electronic absorption spectrum has been interpreted on this basis.

11.2   THREE COORDINATION

Only a few examples of three coordination are known in the solid state, and these are also restricted to the d^{10} configuration. Definite examples are known for Cu(I) and Ag(I), and one example for Pt(O) (Table 11.2). In terms

Table 11.2 Three-Coordinate Compounds

Compound	Bond lengths (pm)	Bond angles
$KCu(CN)_2$	CuN 205, CuC 192(2)	Cu angles 112°, 107°, 134°
$K[Cu_2(CN)_3]3H_2O$		
$(C_6H_5N{=}N{-}NC_6H_5)Cu$		
CuCl(cyclooctatetrene)		
$AgClO_4$(cyclohexylbenzene)	Trigonal pyramid	
$Ag(CN)_3$	AgN 221, 225(3)	
$Pt(Ph_3P)_3$	PtP 226	Trigonal planar

of what has been said in Section 11.1, it is to be expected that Cu(I) and Ag(I) would form three-coordinate compounds more readily than Au(I).

The stoichiometry MX_3 is not a good guide to three coordination, owing to the possibility of polymerization by either bridging ligands or the formation of metal-metal bonds. For example, $K_2[CuCl_3]$ and $Cs_2[AgCl_3]$ contain four-coordinate Cu and Ag through chlorine bridging.

The two Cu(I) cyanides (Table 11.2) have chain structures in which the metal achieves threefold coordination. The short Cu—Cu distance in the complex $Cu[PhN_3Ph]$ probably results from the steric requirements of the ligand rather than a specific Cu—Cu bond. However, owing to this effect, the copper attains the three coordination.

11.3 FOUR COORDINATION

Four coordination is the second most common coordination number for the transition metals; it can occur as a square planar or tetrahedral stereochemistry. The range of metal ions that form compounds with these shapes are listed in Tables 11.3 and 11.4.

Square Planar Stereochemistry

The square planar stereochemistry is relatively common for the transition metals with the configuration d^n, where $9 \geqslant n > 6$ (see Table 11.3). The shape is closely related to the two dominant stereochemistries of the transition metals—viz. octahedral and tetrahedral. Removal of two

Table 11.3   Square Planar

Group		d^9	d^8	d^7	d^6	d^5	d^4	d^3
							Configuration	
VI							Cr(II)	Cr(III)
VII			Mn($-$II)	Mn($-$I)		Mn(II)		
VIII	*a*				Fe(II)			
	b				Co(II)			
				Rh(I)*	Rh(II)			
				Ir(I)*	Ir(II)			
	c	Ni($-$I)	Ni(I)	Ni(II)*				
				Pd(II)*				
				Pt(II)*				
I			Cu(II)*	Cu(III)				
			Ag(II)	Ag(III)				
			Au(II)?	Au(III)*				

* Principal examples.

trans ligands from an octahedron, or bending of the $M-L$ bonds of a tetrahedron, both give rise to a square plane (Fig. 11.2). Compounds exist which have stereochemistries intermediate between octahedral and square planar (tetragonal distortion of octahedron), and tetrahedral and square planar.

Bonding. Simple molecular orbital theory applied to a square planar complex (D_{4h} symmetry) leads to the energy level diagram given in Fig. 11.3. We find two important results: the ordering of the d orbitals is probably $d_{x^2-y^2} \gg d_{xy} > d_{z^2} > d_{xz}, d_{yz}^{1}$; the energy separation of the $d_{x^2-y^2}$ and d_{xy} orbitals is large. The d^8 configuration is prominent in square planar complexes, the last point being the reason why the configuration is ideally suited for spin-paired square planar complexes.

The molecular orbital energy diagram (Fig. 11.3) is applicable to complexes formed with σ-bonding ligands (e.g., $PtCl_4^{2-}$). However, if metal-ligand π bonding can also occur [as in $Pt(CN)_4^{2-}$], further molecular orbitals will be included in the diagram, but the basic d orbital energy sequence

Table 11.4   Tetrahedral

Group	d^0	d^1	d^2	d^3	d^4	d^5	d^6	d^7	d^8	d^9	d^{10}
					Configuration						
IV	Ti(IV)* Zr(IV)*										
V	V(V)	V(IV)	V(III)								
VI	Cr(VI) Mo(VI) W(VI)	Cr(V)	Cr(IV)	Cr(III)?							
VII	Mn(VII) Tc(VII)* Re(VII)*	Mn(VI)	Mn(V)			Mn(II)					Mn(−III)
VIII a	Ru(VIII) Os(VIII)	Ru(VII) Os(VII)	Fe(VI) Ru(VI)	Fe(V)		Fe(III)	Fe(II)				Fe(−II) Ru(−II) Os(−II)
b							Co(III)	Co(II)*	Co(I)	Co(O)	Co(−I) Rh(−I) Ir(−I)
c									Ni(II)*	Ni(I)	Ni(O) Pd(O) Pt(O)
											Ni(−I)?
I										Cu(II)	Cu(I)* Ag(I)* Au(I)

* Principal examples.

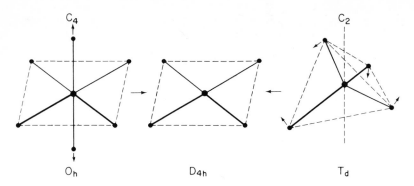

Fig. 11.2 Relationship between an octahedron (O_h), square plane (D_{4h}), and tetrahedron (T_d).

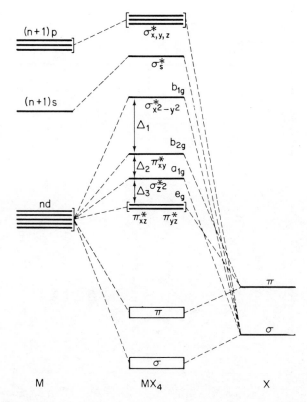

Fig. 11.3 Molecular orbital energy diagram for a square planar complex.

is not affected. From the MO diagram it is clear that three d-d type electronic transitions are possible:

$$(b_{2g})^2 \rightarrow (b_{2g})^1(b_{1g})^1, \quad {}^1A_{1g} \rightarrow {}^1A_{2g}, \quad \Delta_1$$

$$(a_{1g})^2 \rightarrow (a_{1g})^1(b_{1g})^1, \quad {}^1A_{1g} \rightarrow {}^1B_{1g}, \quad \Delta_1 + \Delta_2$$

$$(e_g)^4 \rightarrow (e_g)^3(b_{1g})^1, \quad {}^1A_{1g} \rightarrow {}^1E_g, \quad \Delta_1 + \Delta_2 + \Delta_3$$

The electronic spectra of square planar complexes of Pt(II), Pd(II), Ni(II), and Au(III) can be assigned on this basis. Estimates of the energies Δ_1, Δ_2, and Δ_3 can be made. Table 11.5 gives some typical values.

Table 11.5 Splitting Parameters for Square Planar Complexes

Complex ion	Δ_1	Δ_2	Δ_3 (cm^{-1})
$[PdCl_4]^{2-}$	19,150	6,200	1,450
$[PtBr_4]^{2-}$	22,150	6,000	3,550
$[Ni(CN)_4]^{2-}$	24,950	9,900	650
$[Pd(CN)_4]^{2-}$	>30,000	10,800	50
$[Pt(CN)_4]^{2-}$	>30,000	12,600	4,140
$[Au(CN)_4]^{2-}$	33,410	10,620	

For the d^7, d^8, and d^9 configurations, the orbital occupancy is $(e_g)^4(a_{1g})^2(b_{2g})^1$, ${}^2B_{2g}$; $(e_g)^4(a_{1g})^2(b_{2g})^2$, ${}^1A_{1g}$; and $(e_g)^4(a_{1g})^2(b_{2g})^2(b_{1g})^1$, ${}^2B_{1g}$, respectively. If the compounds are square planar and the theory is correct, the magnetic moments should correspond to the predicted ground states. This is found to be the case for Co(II) d^7, Ni(II), Pd(II), Pt(II) d^8, and Cu(II) d^9. For configurations d^n, where $n < 7$, square planar complexes are restricted to phthalocyanine complexes of Cr(II) d^4(?), Cr(III) d^3, Mn(II) d^5, and Fe(II) d^6. The stereochemistry is a requirement of the ligand and not the metal. In a number of these cases, the magnetic properties are less readily explained (Table 11.6), partly because of the problem of determining the order of orbital energies and how best to fill them, and partly because of the uncertainty of the contribution to the magnetism of other electronic states.

The Square Planar Problem. It is not always easy to decide when a complex is a true square plane rather than a strong tetragonally distorted octahedron. Until recently, all diamagnetic complexes with the d^8 configuration have been assumed to be square planar. However, this need not be the case, provided that the parameter Δ_1 (Figs. 11.3 and 11.4) for a distorted octahedral complex is greater than the interelectronic repulsion term. Some optimum distortion exists, below which paramagnetism will occur and above which

Table 11.6   Phthalocyanine Complexes

Configuration	Complex	Unpaired electrons	S	μ_{obs}	Possible ground state
d^4	[Cr(II)(phth)]	4	2	3.5	Cr$\cdots$Cr interaction is probable
d^3	[Cr(III)(phth)]$^+$OH$^-$	3	$\frac{3}{2}$	4.03	$(b_{2g})^1(e_g)^1(a_{1g})^1$
d^5	[Mn(II)(phth)]	3	$\frac{3}{2}$	4.34	$(b_{2g})^2(e_g)^2(a_{1g})^1$
d^6	[Fe(II)(phth)]	2	1	3.85	$(b_{2g})^2(e_g)^3(a_{1g})^1$
d^7	[Co(II)(phth)]	1	$\frac{1}{2}$	2.12	$(e_g)^4(b_{2g})^2(a_{1g})^1$
d^8	[Ni(II)(phth)]	0	0	0	$(e_g)^4(b_{2g})^2(a_{1g})^2$
d^9	[Cu(II)(phth)]	1	$\frac{1}{2}$	1.73	$(e_g)^4(b_{2g})^2(a_{1g})^2(b_{1g})^1$

spin pairing will be achieved. Since $10\,Dq \propto 1/r^6$, it is possible to have a tetragonally distorted octahedral stereochemistry where the ligand field imposed is essentially that of a square plane. Because of the two-dimensional nature of square planar compounds, other groups may occur above and below the plane. This can be achieved in a number of ways, such as nonbonding groups blocking the positions above and below or metal-metal interaction or metal-ligand interaction or metal-ion interaction, and, in solution, metal-solvent interaction (see p. 210).

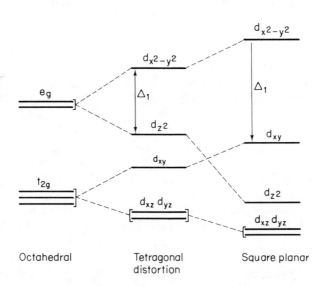

Fig. 11.4   Splitting of *d* orbitals in a tetragonally distorted octahedral complex and in a square planar complex.

A number of square planar complexes pack in the crystalline state, so that the next nearest neighbors of the metal ions are the metal ions of neighboring molecules (Fig. 11.5). In a number of cases, the distances have been

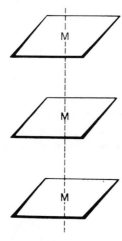

Fig. 11.5 Stacking of square planar complexes.

determined, some of which are given in Table 11.7. The metallic radius for the metal atoms listed in the table are Ni 125, Pd 137, Pt 139, and Cu 128 pm. The metal-metal distances given in Table 11.7 are greater than the distances $2r$ or $r_A + r_B$. However, spectroscopic evidence indicates that in some

Table 11.7   Metal–Metal Distance in Square Planar Complexes

Compound	M—M distance (pm)
$Ni(dmg)_2$	325
$Pd(dmg)_2$	326
$Pt(dmg)_2$	323
$\alpha Ni(N\text{-}MeSa)_2$	330
$\alpha Cu(N\text{-}MeSa)_2$	333
$\alpha Pd(N\text{-}MeSa)_2$	328
$Pt(NH_3)_4PtCl_4$	323, 324
$Pd(NH_3)_4PtCl_4$	325
$Cu(NH_3)_4PtCl_4$	323, 322
$Pt(MeNH_2)_4PtCl_4$	325
$Pd(NH_3)_4PdCl_4$	325
$Pt(NH_3)_4Pt(SCN)_4$	335
K_2PdCl_4	410
K_2PtCl_4	413

compounds, metal-metal interaction takes place. For example the Magnus green salt $[Pt(NH_3)_4][PtCl_4]$ has a Pt—Pt distance of 323 pm, whereas the pink salt has a distance > 500 pm. The pink color is the combined color of the constituent ions, whereas the green color is associated with the Pt—Pt interaction. When the metal-metal distance is even shorter, as in $[Cu(acetate)_2\cdot H_2O]_2$ (Cu—Cu = 264 pm), the interaction can be related to a change in the magnetic susceptibility.

The tendency to achieve sixfold coordination around the metal ion is even more pronounced for square planar Cu(II) d^9. In a number of cases, considerable distortion of the ligand takes place in an attempt to attain the higher coordination. For example, the bulk of the phthalocyanine ligand in the Cu(II) complex only allows a fifth and sixth L—Cu contact of ~ 340 pm; however, in the case of Cu(II) (benzeneazo-β-napthol)$_2$, the ligand bends in order to give an axial Cu—L distance of 250 pm, while the $L\ldots L$ distance is maintained at 340 pm (Fig. 11.6).

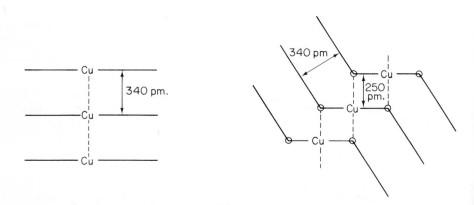

Fig. 11.6   Bending of ligand around copper in order to achieve sixfold coordination of the metal ion.

Square Planar Complexes and Electron Configuration. The d electron configurations that occur with the square planar stereochemistry are the d^8, d^9, and d^7, in decreasing order of abundance (Table 11.3). We can show that from different theoretical points of view these are the configurations most likely to give planar complexes.

In terms of available bonding orbitals, the configurations d^n, where $n > 6$, are more likely to be associated with low coordination numbers, particularly for spin-paired states. The d^8 spin-paired configuration has one vacant d orbital available for bonding, and from a valence bond point of view one can envisage the use of dsp^2 hybrids. The d-p orbital energy separation is important in determining how readily the d and p orbitals mix, especially the mixing in of the third p orbital to give five coordination. For the first row transition metals with the d^8 configuration [Cr($-$II), Mn($-$I), Fe(O), Co(I), Ni(II), and Cu(III)], the tendency toward four-coordinate square planar complexes increases to the right (compared with five coordination) as the d-p energy separation increases. For the dipositive cations [Ni(II), Pd(II), Pt(II)], the tendency increases toward the heavier metals.

We have already pointed out (p. 198) that from MO theory the spin-paired d^8 configuration is ideal for forming square planar complexes. The change from distorted octahedral to square planar stereochemistry corresponding to the configurations $(e_g)^4(a_{1g})^2(b_{2g})^1(b_{1g})^1$ and $(e_g)^4(a_{1g})^2(b_{2g})^2$, respectively, is associated with a loss in exchange energy and increase in interelectronic repulsion energy. But this is compensated for by improved σ bonding and the possibility of strong π bonding in the planar structures. The improved σ bonding results from the use of the $d_{x^2-y^2}$ orbital, which is antibonding in the paramagnetic octahedral complexes. The strong π bond results from the use of the unused p_z orbital, which is not available in an octahedral complex. The incorporation of the $d_{x^2-y^2}$ orbital in the σ bonding and the p_z orbital in the π bonding increases in the order Ni < Pd < Pt. The same explanation can be used for the d^7 spin-paired configuration, but in the case of the d^9 configuration, a strong electronic distortion is probably the main reason for the square planar arrangement.

Another approach can be made by comparing the crystal field stabilization energies (CFSE) for each configuration, remembering, however, that the CFSE is only 10% or less of the total binding energy. Bearing this in mind, we can draw certain broad conclusions from a consideration of CFSE's. The figures given in Table 11.8 demonstrate that the planar stereochemistry is favored over octahedral for the d^8, d^7, and, to a lesser extent, d^9 configurations in a strong field. It turns out that except for d^0, d^{10}, and d^5 (spinfree), the CFSE is always greatest for the square planar shape. But it is only when the difference is greater than 6 Dq that it is reasonable to predict the existence of the planar stereochemistry.

Table 11.9 lists the predictions of the different theoretical approaches.

Ligand Influence on Square Planarity. In addition to certain configurations favoring the square planar arrangement, ligands themselves can have a considerable electronic effect. This is especially important for the first row transition metals, where there is a possibility of both spinfree and spin-paired states, the latter occurring in square planar complexes.

If the p_z orbital is involved in $M - L$ π bonding, then it is unavailable for the bonding of ligands above and below the plane. Therefore, if one selects ligands that will form a π bond with the p_z orbital, it is possible to stabilize the

Table 11.8   Crystal Field Stabilization Energies

	Octahedral		Square planar	
Configuration	Weak	Strong	Weak	Strong
d^0	0	0	0	0
d^1	4	4	5.14	5.14
d^2	8	8	10.28	10.28
d^3	12	12	14.56	14.56
d^4	6	16	12.28	19.70
d^5	0	20	0	24.84
d^6	4	24	5.14	29.12
d^7	8	18	10.28	26.84*
d^8	12	12	14.56	24.56*
d^9	6	6	12.28	12.28*
d^{10}	0	0	0	0

* CFSE favoring square planar stereochemistry.

Table 11.9   Predictions of Theoretical Approaches to Square Planar Complexes

Theory	Predictions
Nine-orbital rule	d^n, where $n \geqslant 7$, especially spin paired
VB theory and d-p energy separation	d^8; Cu^{3+}, Ni^{2+} > other first row metal ions with d^8 configuration; and 3rd row > 2nd row > 1st row
MO theory	d^7, d^8, and d^9 spin-paired; Pt(II) > Pd(II) > Ni(II) based on the use of $d_{x^2-y^2}$ and p_z orbitals
CFSE	d^7, d^8, and possibly d^9 spin paired

planar stereochemistry.[2] Dithiolato- ligands are good examples of this type of ligand, e.g., maleonitriledithiolate (MNT).

$(MNT)^{2-}$

A wide range of square planar complexes are formed with $(MNT)^{2-}$ (Table 11.10). The complexes can be oxidized to $[M(MNT)_2]^-$, but it is not certain whether the metal or one of the ligands is oxidized. The latter seems the more probable because neutral complexes $[M(MNT)_2]^0$ can also be obtained.

Table 11.10   Square Planar Complexes with MNT*

Complex type	Metal	Magnetic moment μ(BM)	d-Electron configuration	Total spin quantum number
$[M(II)(MNT)_2]^{2-}$	Cu	1.85	d^9	$\frac{1}{2}$
	Au	1.86	d^9	$\frac{1}{2}$
	Ni†	Diamag.	d^8	0
	Pd	Diamag.	d^8	0
	Pt	Diamag.	d^8	0
	Co†	2.12	d^7	$\frac{1}{2}$
	Rh	1.91	d^7	$\frac{1}{2}$
$[M(MNT)_2]^-$ ‡	Cu†	Diamag.	d^8	0
	Au	Diamag.	d^8	0
	Ni†	1.20 (solid)	d^7	$\frac{1}{2}$
		1.76 (soln)		
	Pd	Diamag.	d^7	$\frac{1}{2}$
	Pt	1.05 (solid)	d^7	$\frac{1}{2}$
		1.80 (soln)		
	Co	Diamag.(solid)	d^6	0
		2.81 (soln)	d^6	1

*MNT – maleonitriledithiolate.
† Structures have been determined.
‡ Formally M(III) and 2(MNT)$^{2-}$ *or* M(II) and (MNT)$^-$ + (MNT)$^{2-}$.

Similar ligands that give the planar stereochemistry are

, and CN^-

The difference between the cyano- and chloro- ligands highlights the π-bonding property. The cyanides $[M(CN)_4]^{2-}$ (M = Pt, Pd, and Ni) are square planar, just as are the chloro- complex ions $[PtCl_4]^{2-}$ and $[PdCl_4]^{2-}$. However, the absence of π bonding in the chloro- complexes cannot stabilize the spin-paired Ni(II) complex; instead, this leads to the formation of the spinfree tetrahedral anion $[NiCl_4]^{2-}$.

The quadridentate phthalocyanine ligands, by virtue of their own fixed planar nature, force the metal into forming square planar complexes (Table 11.6). The quadrivalent Schiff base ligands are further examples of ligands that form a square planar environment around the metal ion. Complexes formed with such ligands may still become five or six coordinate, inasmuch as the square planar stereochemistry is a steric requirement of the ligands and not necessarily a bonding requirement as for the dithiolato ligands.

Ligand bulk can also affect the stereochemistry. For example, the bulky mesitylene group in the complex *trans*-Co(mesityl)$_2$(PEt$_2$Ph)$_2$ determines the shape in preventing other ligands approaching the metal ion. The mesostilbenediamine complex of Ni(II) (Lifschitz complex) is another example of a complex where the ligand blocks the other coordination sites.

A different type of steric effect is observed with ligands having donor arrangements as follows:

$$
\begin{array}{ccc}
\begin{array}{c}
-X \\
\quad \diagdown \\
\qquad M \\
\quad \diagup \\
-N \\
\;| \\
\;R
\end{array}
& \text{and} &
\begin{array}{c}
R \\
\diagup \\
X \\
\;\diagdown \\
\qquad M \\
\;\diagup \\
X \\
\diagdown \\
R
\end{array}
\end{array}
$$

as in ketoaminates and dipyrromethene complexes. In the first case, square planar complexes are formed when the group R is small, but as the size increases, steric influence produces the tetrahedral stereochemistry. The bidentate monoanion dipyrromethene ligands can be formed with a variety

of substituents in the 2, 10 positions. In complexes of the type $[M(II)L_2]$, the steric interaction of these substituents produces near tetrahedral complexes for Ni(II), Co(II), Zn(II), Cd(II), and Hg(II); but for Pd(II), the square planar stereochemical requirement of the metal dominates, and the highly conjugated ligand buckles at carbons 5 and 7 to overcome the steric interactions of the C$_2$ and C$_{10}$ substituents.

Isomerism. Square planar complexes of the type ML_2L_2' and $ML_2L'L''$, where L, L', and L'' are monodentate ligands, lead to cis and trans geometric isomerism (Fig. 11.7). Isomerism of this type has been studied in detail for divalent platinum complexes. Geometric isomerism for four-coordinate compounds distinguishes square planar from tetrahedral complexes, because geometric isomerism does not occur for the latter.

The "trans effect" has been discussed in Chapter 7 (Section 7.5), and the trans influence of ligands on chemical reactivity has also been discussed.

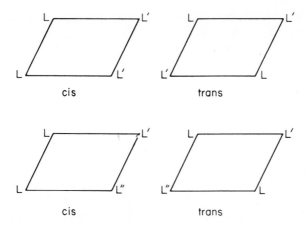

Fig. 11.7 Cis and trans geometric isomers.

A detailed discussion of the trans effect can be obtained elsewhere,[3] but another point of stereochemical interest is the structural implication of the trans effect. Lability of certain ligands implies weaker bonds; this is borne out from bond lengths in Pt(II) complexes (Table 11.11). The Pt-halogen

Table 11.11   Trans Effect in Platinum Complexes

Compound	Trans group	Bond length Pt-halogen (pm)
$K[Pt(NH_3)Cl_3]$	Cl	235
	NH_3	232
$K[Pt(NH_3)Br_3]$	Br	270
	NH_3	242
$K[Pt(C_2H_4)Cl_3]H_2O$	Cl	232
	C_2H_4	242
$K[Pt(C_2H_4)Br_3]H_2O$	Br	242
	C_2H_4	250
t-$[Pt(Et_3P)_2HBr]$	H	256
t-$[Pt(EtPh_2P)_2HCl]$	H	242
c-$[Pt(Me_3P)_2Cl_2]$	Me_3P	238
t-$[Pt(Et_3P)_2Cl_2]$	Cl	229

bond trans to a ligand with a greater trans effect than the halogen is lengthened to greater than the sum of the covalent radii, 230 pm for Pt—Cl and 243 pm for Pt—Br. Also, the bond lengthening is greatest for ligands with the highest trans effect.

Nickel (II). The stereochemical changes that take place for Nickel(II), d^8, spin-paired planar complexes will now be discussed in more detail. In solution, the following equilibrium may occur in sufficiently coordinating solvents:

$$NiL_4 + 2S \rightleftharpoons \textit{trans-}NiL_4S_2$$

diamagnetic,	paramagnetic
square planar	octahedral

The square planar complex has coordinated two further ligands, which normally reduces the splitting energy Δ_1 (Fig. 11.3) so that NiL_4S_2 is paramagnetic. Since the reaction is an equilibrium, the magnetism of the solution will lie somewhere between that for zero and that for two unpaired electrons. Sometimes the reaction proceeds completely to the right-hand side, as for N-alkylsalicylaldiminato nickel(II) complexes in pyridine. Nickel compounds of substituted ethylene diamines called Lifschitz complexes—e.g.,

$$\begin{array}{c} R \qquad\qquad H_2 \quad H_2 \qquad\qquad R \\ \diagdown \qquad\qquad\qquad\qquad\qquad\qquad \diagup \\ CH-N \qquad\qquad N-CH \\ | \qquad \diagdown \quad \diagup \qquad | \\ \qquad\qquad Ni \qquad\qquad \\ | \qquad \diagup \quad \diagdown \qquad | \\ CH-N \qquad\qquad N-CH \\ \diagup \qquad H_2 \quad H_2 \qquad\qquad \diagdown \\ R \qquad\qquad\qquad\qquad\qquad\qquad\qquad R \end{array}$$

also display the square planar octahedral equilibrium. Yellow complexes are square planar whereas blue ones are octahedral. Conversion between the two can be effected by temperature or change of solvent. The effect of temperature changes the position of equilibrium. In the case where an octahedral compound is formed (and where Δ_1 is close to the interelectronic repulsion energy), changes in the population of the spinfree and spin-paired states will be produced by temperature changes. The complex Ni(NN′-diethylthiourea)$_4$Cl$_2$ is an example of this situation. The crystalline form of *bis*(mesostilbenediamine)Ni(II)dichloroacetate ($R = C_6H_5$ in preceding formula) contains both square planar and octahedral species, whereas in the latter case the fifth and sixth positions are filled by oxygen of the dichloroacetate ion.

The acetylacetonates of divalent nickel provide another example of stereochemical changes. Six coordination is achieved for anhydrous Ni(acac)$_2$ by polymerization (Fig. 11.8), but this can be prevented by introducing bulky substituents at C_1 and C_5 of the acetylacetone. The results are summarized in Table 11.12.

The third stereochemical change occurs from square planar to tetrahedral. In addition to the examples already given, the complexes of the type

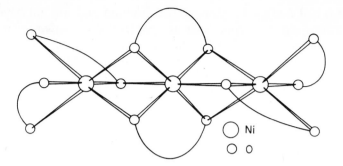

Fig. 11.8 Structure of $[Ni(acetylacetone)_2]_3$. Curved lines correspond to the organic backbone of the acetylacetone.

Table 11.12 Ni(II) Acetylacetonates

$$\begin{bmatrix} \begin{array}{c} CR_3 \\ | \\ C-O \\ \parallel \\ CH \qquad Ni \\ \diagdown \\ C=O \\ | \\ CR_3 \end{array} \end{bmatrix}_2$$

$R_3 = H_3$	$H(CH_3)_2$	$(CH_3)_3$
Trimeric octahedral spin free	Partial polymerization; magnetism is a function of temperature and solvent	Monomeric square planar spin paired

NiL_2X_2, where L is a phosphine and X is a halogen, display this isomerism. It is found that the trend toward square planarity occurs in the orders

$$Ph_3P < Ph_2EtP < PhEt_2P < Et_3P$$

and

$$I^- < Br^- < Cl^- < SCN^-$$

The trend suggests that a weak ligand field favors the tetrahedral stereochemistry. In some cases, both shapes can be obtained with the same ligand. Crystalline $\{Ni[(C_6H_5CH_2)(C_6H_5)_2P]_2Br_2\}$ contains both tetrahedral and square planar species.

Rhodium(I) and Iridium(I). Monovalent rhodium and iridium have the d^8 configuration, and their complexes are predominantly square planar. The rhodium compound $[Rh(CO)_2Cl]_2$ has the structure given in Fig. 11.9. The compound is a useful starting material for the preparation of a number of

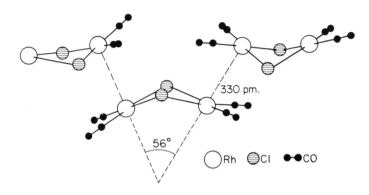

Fig. 11.9   Structure of $[Rh(CO)_2Cl]_2$.

four-coordinate Rh(I) compounds, which are presumably square planar. Another important d^8 complex is the iridium compound $Ir(CO)Cl(Ph_3P)_2$. The five-coordinate dioxygen complex $(O_2)Ir(CO)Cl(Ph_3P)_2$ shows random distribution of the CO and Cl ligands, suggesting that the oxygen attacks from both sides of the square planar molecule.

Tetrahedral Stereochemistry

Electron Configuration. The transition metals that form tetrahedral species are listed in Table 11.4. It is distinctly clear (a) that the majority of tetrahedral compounds occur for the d^0 and d^{10} configurations, and (b) that tetrahedral compounds for metal ions with the remaining configurations, especially d^3 to d^9, are restricted to the first row transition metals. The limited range of tetrahedral compounds is a result of the arrangement of the four ligands, which, although giving minimum ligand repulsions, impose a weak ligand field. The relation of the tetrahedral ligands to the lobes of the metal p or d orbitals (Fig. 11.10) is such that none of the orbitals is lying directly along the $M-L$ bonds. Therefore, the potential applied by the four ligands is less effective than in the case of an octahedral or square planar field. An additional factor in comparing tetrahedral and octahedral structures is that in the former case the metal ion is surrounded with two ligands less. The overall tetrahedral splitting parameter $Dq_{(t)}$ is $\frac{4}{9}$ the octahedral parameter $Dq_{(O)}$,

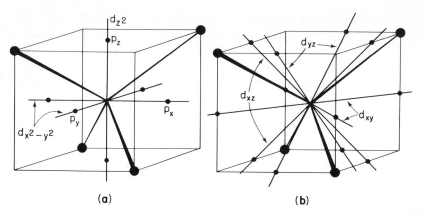

Fig. 11.10 The relation of tetrahedral ligands to metal p and d orbitals: (a) p_x, p_y, p_z, d_{z^2}, and $d_{x^2-y^2}$; (b) d_{xy}, d_{yz}, d_{xz}.

which means that first row transition metals always form spinfree tetrahedral complexes. However, for the heavy transition metals where the inter-electronic repulsion energy is less, it is possible that a spin-paired tetrahedral complex may form; as yet, no examples are known. The d^4 electron configuration is the most likely candidate, i.e., Mo(II), W(II), Tc(III), Re(III), Ru(IV), and Os(IV).

The crystal field stabilization energies (CFSE) for each d configuration for tetrahedral and weak octahedral fields are compared in Table 11.13. On

Table 11.13 Crystal Field Stabilization Energies

Configuration	Spinfree octahedral CFSE $Dq_{(O)}$	Tetrahedral CFSE (Dq) in units of $Dq_{(O)}$
d^0, d^5, d^{10}	0	0
d^1, d^6	4	2.67
d^2, d^7	8	5.34
d^3, d^8	12	3.56
d^4, d^9	6	1.78

the basis of the CFSE, it is clear that only the d^0, d^{10}, and d^5 (spinfree) configurations are likely to form tetrahedral structures as readily as octahedral. This conclusion is borne out for the d^0 and d^{10} configuration (Table 11.4). For the remaining d^n configurations (using CFSE), the expected order of decreasing ability to form tetrahedral compounds is Fe(II) > Co(II) > Cu(II) ≫ Ni(II), Co(I). Proponents of CFSE have suggested that it would not be possible to form tetrahedral Ni(II) complexes. In fact, this is not so. With both the right ligand and experimental conditions {i.e., in the absence of water, which forms the stable octahedral cation $[Ni(H_2O)_6]^{2+}$}, tetrahedral complexes of Ni(II) have been formed (Table 11.14). Nyholm[4] has outlined a number of reasons why the CFSE comparison of octahedral

Table 11.14   Tetrahedral Nickel(II) Complexes

$[NiX_4]^{2-}$ (X = Cl, Br, I)
$[Ni(dpm)_2]$*
$[NiCl_2(Ph_3P)_2]$
$[NiBr_2(Ph_3AsO)_2]$
Ni^{2+} in $NiCr_2O_4$

* dpm = Dipyrromethene.

and tetrahedral structures is invalid in this case. First, the energy under consideration is only 5 to 10% of the total binding energy. Second, the crystal field approach considers all $M—L$ interactions as ionic, ignoring the possibility of covalent bonding. Third, the comparison assumes that the ligand involved in the tetrahedral and octahedral complexes is the same in each case, and that the $M—L$ bond length is identical in the two stereochemistries. Fourth, steric effects are ignored. All in all, the CFSE approach must be applied with considerable caution. The approach is informative in indicating which configuration will most readily form tetrahedral complexes. However, this is only relevant for the situation where a related series of tetrahedral complexes $ML_4^{n\pm}$ are considered, (L and n are kept constant and M varies). Otherwise, comparisons between one stereochemistry and another are of doubtful usefulness.

Ligand Influence. Some general points can be made concerning the nature of the ligands that favor the tetrahedral stereochemistry.

(a) Steric effects are important, especially when steric interactions prevent the formation of square planar or octahedral complexes. In the case of the dipyrromethene complexes mentioned on p. 208 the copper(II) complex is a distorted tetrahedron and the angle between the planes containing the two ligands is 66° (90° for a regular tetrahedron). The actual shape taken up is a compromise between the square planar requirement of the metal ion and the steric requirements of the ligand. The polymerization of iron(II), cobalt(II), and nickel(II) acetylacetonates in order to achieve octahedral coordination around the metal has been mentioned (compare Figs. 11.8 and 11.11). Replacement of the terminal $—CH_3$'s with the bulky $—C(CH_3)_3$ groups, to give dipivaloylmethane, produces tetrahedral com-

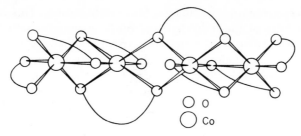

Fig. 11.11 Structure of $[Co(acetylacetone)_2]_4$. The curved lines represent the organic backbone of the acetylacetone.

pounds with iron(II) and cobalt(II). The steric interaction of the $-C(CH_3)_3$ groups prevents polymerization.

(b) A negatively charged ligand is more likely to form tetrahedral complexes due to interligand repulsion, which is minimized in a tetrahedral environment (compared with square planar).

(c) Since the tetrahedral ligand field energy is small, ligands that favor the tetrahedral stereochemistry are "weak," i.e., do not produce spin pairing. Similarly, ligands that are not π acceptors are associated with the tetrahedral stereochemistry. A metal ion in a tetrahedral environment can participate in π bonding, but, like the σ bonding, it is not as effective as in the octahedral and planar stereochemistries. Therefore, it is not surprising that $[NiCl_4]^{2-}$ is tetrahedral whereas $[Ni(CN)_4]^{2-}$ is square planar.

Transition Metal Halogeno-Complexes. The complex anions $[M(II)X_4]^{2-}$ (M = Mn, Fe, Co, Ni, Cu, and Zn) and $[M(III)X_4]^-$ (M = V and Fe), where $X = Cl^-$, Br^-, I^-, and SCN^-, are tetrahedral with differing degrees of distortion. The metal ions absent from the series, V(II), Cr(III), Cr(II), and Mn(III), have the d^3 and d^4 configurations where the spinfree arrangement in a tetrahedral environment would lead to considerable distortion and destabilization of the stereochemistry.

If the complex anions are considered from an ionic point of view, then the radius ration $r_{M^{n+}}/r_{X^-}$ for the chloro- complexes (Table 11.15) lies

Table 11.15 Radius Ratio for $M(II)$ and $M(III)$ Halogeno- Complexes*

	Mn	Fe	Co	Ni	Cu	Zn
$r_{M^{2+}}$ (pm)	80	74	72	69	72	74
$r_{M^{2+}}/r_{Cl^-}$	0.44	0.41	0.40	0.38	0.40	0.41
$r_{M^{3+}}$ (pm)	60	61				
$r_{M^{3+}}/r_{Cl^-}$	0.33	0.34				

* The ratio r_M/r_X for four coordination is 0.225–0.414 and for six coordination is 0.414–0.732.

within, or very close to, the range for four coordination. This is even more apparent for bromo- and iodo- ligands. Such an analysis is crude, but does suggest that the reason for the limitation to four halogens is the relative size of the metal ion and ligand. The smaller F^- ion, however, readily forms hexafluoro- anions. Therefore, the size restricts the coordination number to four and the nature of the halogen ligand (see section under Ligand Influence) favors the tetrahedral stereochemistry. The large metal ions of the second and third row transition metals invariably give rise to six-coordinate (or greater) halogeno- anions.

One further aspect of the tetrahalogeno- anions of the divalent ions Mn to Zn is that, except for Co and Zn, which form these species most readily, they are isolated with large cations such as R_4As^+, $R_4P^+ \cdot R_4N^+$, and Cs^+ and Rb^+ [for Cu(II)], which form relatively insoluble salts. The large organic cations will also protect the tetrahedral species (in the solid state) from atmospheric moisture and conversion to more stable octahedral species.

Tetrahedral Cobalt(II) Complexes. A number of tetrahedral cobalt(II) complexes are known (Table 11.16), and in addition to X-ray structural work

Table 11.16 Tetrahedral Cobalt(II) Complexes

Co(II) in Co_3O_4
$[CoX_4]^{2-}$ $(X = Cl^-, Br^-, I^-, SCN^-, N_3^-, OH^-)$
Co(bidentate)$_2$(bidentate $= N$-alkylsalicylaldimine, dipyrromethenes)
$(Ph_4As)_2Co(trifluoroacetate)_4$
Co(dipivaloylmethane)$_2$
$CoCl_2(p\text{-toluidine})_2$
$CoI_2(p\text{-toluidine})_2$
$CoCl_4$ in $Cs_3[CoCl_5]$

both electronic spectra and magnetism are valuable in determining the stereochemistry. There are three spin-allowed transitions for tetrahedral cobalt(II), viz., $^4A_2 \rightarrow {}^4T_2$ (3000–5000 cm^{-1}), $^4A_2 \rightarrow {}^4T_1(F)$ (4000–8000 cm^{-1}), and $^4A_2 \rightarrow {}^4T_1(P)$ (14,000–17,000 cm^{-1}) (Fig. 11.12). Normally, the last two transitions only are observed, and the one of highest energy has a molar extinction coefficient in the range of from 300 to 500. The high value is typical of a tetrahedral complex.

The ground term for tetrahedral Co(II) is $^4A_2$, which is free of spin-orbit coupling. However, owing to mixing in of a higher energy T term, the magnetic moment (μ_{eff}) is greater (4.4–4.8 B.M.) than the spin-only value

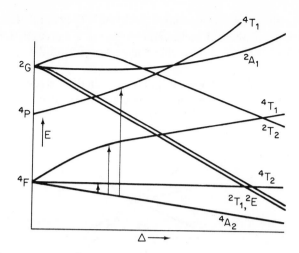

Fig. 11.12 Ligand field energy diagram for the d^7 configuration in a tetrahedral field.

(μ_{SO}) of 3.89 B.M. The moment is described by the equation

$$\mu_{eff} = \mu_{SO}\left(1 - \frac{4\lambda}{10Dq_t}\right)$$

(λ is the spin-orbit coupling constant and is negative). Therefore, a direct correlation exists between μ_{eff} and $10Dq_t$, as illustrated in the data in Table

Table 11.17 Tetrahedral Cobalt(II) Complexes

Complex ion	$[CoI_4]^{2-}$	$[CoBr_4]^{2-}$	$[CoCl_4]^{2-}$	$[Co(NCS)_4]^{2-}$
μ_{obs}	4.77	4.69	4.59	4.40
Approximate $10Dq_t$ cm^{-1}	2600	2800	3100	4500

11.17.[5] This assumes that λ is constant for the various complexes, which is saying, in effect, that the covalency of the metal-ligand bond is the same for each complex.

Tetrahedral Nickel(II) Complexes. The spectral and magnetic characteristics of tetrahedral nickel(II) are similar to those of cobalt(II). Three spin-allowed bands are predicted, viz., $^3T_1(F) \rightarrow {}^3T_2$ (not observed), $^3T_1(F) \rightarrow {}^3A_2$ (7000–8000 cm^{-1}), and $^3T_1(F) \rightarrow {}^3T_1(P)$ ($\sim$15,000 cm^{-1}). Inasmuch as there is spin-orbit coupling associated with the $^3T_1$ ground term, the magnetic moments of tetrahedral nickel(II) complexes are greater than the spin-only value of 2.83 B.M., and lie within the range of from 3.5 to 4.2 B.M.

Tetrahedral Copper Complexes. Regular tetrahedral structures have not been found for copper(II). The nearest is when the two planes indicated in Fig. 11.13 are at 66° to each other (90° for a regular tetrahedron). The

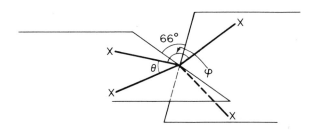

Fig. 11.13   Angle between the two symmetry planes of distorted tetrahedral copper complexes.

stereochemistry of the CuX_4^{2-} anions ($X = Cl$, Br) is that of a squashed tetrahedron, where $90° < \theta < 109.5° < \varphi < 180°$ (Fig. 11.13). Both a Jahn-Teller type distortion for the d^9 configuration and spin-orbit coupling associated with the ground term ($^2T_2$) will influence the stereochemistry. Together, these effects make it difficult to predict the type of distortion. On the other hand, monovalent copper normally exists in a tetrahedral environment, whereas Ag(I) and Au(I) form very few tetrahedral compounds. The copper atoms in the tetrameric complexes of Cu(I) (Table 11.18) are in a

Table 11.18   Tetrahedral Copper(II) and (I) Complexes

	Compound	Comments
Cu(II)	Cs_2CuCl_4	CuCl 222 pm ClCuCl 118°, 104°
	Cs_2CuBr_4	Distortion more marked
	$CuCr_2O_4$	Cu(II) tetrahedral holes
	$Cu(C_3N_2H_3)_2$	Forced by imidazole ligand
	$Cu(dpm)_2$	Forced by ligand
Cu(I)	$[Me_3AsCuI]_4$	CuCu 260 pm
	$[Cu(S_2CNC_2H_5)]_4$	CuCu 276, 266 pm
	$[C_6H_5N_2]Cu_2Br_3$	CuBr 245 pm edge bond, and 257 pm
		center bond. $[Cu_2Br_3]_\infty^-$ crinkled ribbon
	$(Thiourea)_3CuCl$	Tetrahedral Cu
	$[Cu(CNMe)_2I]_n$	Tetrahedral Cu
	$[Cu(II)(NH_3)_4][Cu(I)I_2]_2$	Tetrahedral Cu(I)

tetrahedral environment as well as being arranged tetrahedrally themselves. Direct Cu—Cu bonding also occurs in these compounds.

Oxyanions. Finally, we must mention the discrete tetrahedral anions MO_4^{n-}, where $n = 1, 2,$ or 3, found for metal ions in high oxidation states, on the left of the transition metal series, and with the configurations d^0, d^1, and d^2. Evidence suggests that these are regular tetrahedral species.

11.4   FIVE COORDINATION

The two ideal shapes associated with five coordination are the trigonal bipyramid (TBP) and square pyramid (SP).[6,7] Clearcut information cannot be given as to which structure will form in any given situation, especially for mixed ligand complexes. Often, minor factors can influence the shape assumed by the molecule. For example, the two five-coordinate Ir(I) complexes

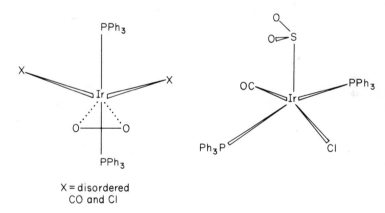

X = disordered
CO and Cl

Fig. 11.14   Structures of [Ir(Ph₃P)₂COCl(O₂)] and [Ir(Ph₃P)₂-COCl(SO₂)].

[Ir(Ph₃P)₂COCl(SO₂)] and [Ir(Ph₃P)₂COCl(O₂)] are square pyramidal and trigonal bipyramidal, respectively (assuming the sideways bonding of the O_2 ligand corresponds to one coordination site). The reason for the difference in the two shapes is probably that the O_2 ligand is more easily accommodated in the trigonal plane of a TBP than in any other position of either shape (Fig. 11.14).

Factors Influencing the Stereochemistry

Electrostatic. If only repulsion of charges is considered, the TBP is favored over the SP, but only slightly favored over a distorted SP, where the apical-base bond angle is around 104° (Fig. 11.15).

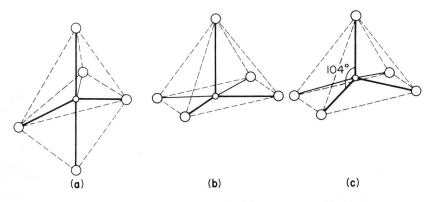

Fig. 11.15   (a) Trigonal bipyramid, (b) square pyramid, (c) distorted square pyramid.

Crystal Field Stabilization Energy. From the point of view of CFSE, the SP is favored over the TBP (Table 11.19). The configurations where the difference is least are d^4, d^6 (spinfree), and d^9 [other than d^0, d^5 (spinfree) and d^{10}, of course]. The data in Table 11.19 are pictured graphically in Fig. 11.16.

Bonding. Regarding the σ- and π-bonding potentials within the two stereochemistries, we may have little to choose. But it has been suggested that if π bonding is important, it is likely to be stronger in an ideal square pyramidal structure.

Ligand Influence. The dominant influence on the stereochemistry is most likely the steric requirements of the ligands, especially for polydentate ligands. For example, a number of complexes with Schiff base ligands have a square planar arrangement of the four donor atoms; if one more ligand is added, the result will be a SP structure. The bonding requirements of the multiple bonded VO group in tetravalent VOL_4 complexes determine the SP stereochemistry. If the shape was TBP, the oxygen would have more competition from the other ligands for the metal's bonding orbitals. The effect of such competition is observed in the decrease in the infrared stretching frequency

Table 11.19 Crystal Field Stabilization Energies

	Weak field		Strong field	
	TBP	SP	TBP	SP
d^0	0	0	0	0
d^1	2.72	4.57	2.72	4.57
d^2	5.44	9.14	5.44	9.14
d^3	6.26	10.00	6.26	10.00
d^4	7.08	9.14	7.08	9.14
d^5	0	0	9.80	13.71
d^6	2.72	4.57	12.52	18.28
d^7	5.44	9.14	13.34	19.14
d^8	6.26	10.00	14.16	18.28
d^9	7.08	9.14	14.16	9.14
d^{10}	0	0	0	0

TBP		SP	
$a'_1(d_{z^2})$	—— 7.08	$b_1(d_{x^2-y^2})$	—— 9.14
	↑ Δ_2 ↓		↑ Δ ↓
$e'(d_{x^2-y^2}, d_{xy})$	══ −0.82	$a_1(d_{z^2})$	—— +0.86
	↑ Δ_1 ↓		
$e''(d_{xz}, d_{yz})$	══ −2.72	$b_2(d_{xy})$	—— −0.86
		$e(d_{xz} d_{yz})$	══ −4.57

of the VO bond when a sixth ligand coordinates trans to the oxygen. A similar situation applies to the five-coordinate oxo- and nitrido- complexes of rhenium(V).

Solid State Effects. Crystal lattice and packing effects may be the reason for the TBP stereochemistry of $[CuCl_5]^{3-}$, whereas most five-coordinate Cu(II) compounds are square pyramidal.

It therefore appears from structural information available (Table 11.20) that (a) if the ligands are monodentate and all the same, or nearly so, and (b) if polydentate ligands do not impose steric constraints, and (c) if the bonding can be considered as largely electrostatic, then the TBP is favored over the SP. However, (a) if steric requirements of polydentate ligands are important, and (b) if π bonding is possible, then the SP is favored over the

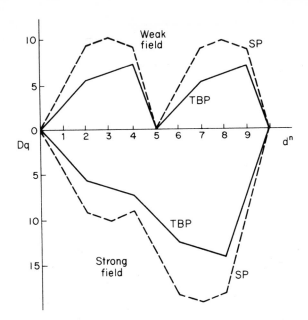

Fig. 11.16 Crystal field stabilization energy plotted against d^n for a trigonal bipyramid and a square pyramid.

TBP. The reader will readily think of a number of exceptions to these generalizations. However, this will only help to emphasize the need for caution in speculation on the possible stereochemistry of five-coordinate compounds. The empirical formula is not a safe guide either. For example, $MoCl_5$ has a trigonal bipyramidal shape in the gaseous state but is octahedral in the solid state, owing to dimerization.

Distortions from a Regular Stereochemistry

The apical bonds that are generally long for the TBP compounds of the nontransitional elements are not necessarily the long ones for the transition elements. The reverse often occurs—e.g., $[Ni(CN)_5]^{3-}$, NiC_{ap} 180 pm, NiC_{eq} 188–200 pm; $[CuCl_5]^{3-}$, $CuCl_{ap}$ 230 pm, $CuCl_{eq}$ 239 pm; and $[Co(CH_3CN)_5]^+$, CoN_{ap} 184 pm, CoN_{eq} 188 pm. The high-energy d_{z^2} orbital (in fact antibonding) of the metal ion in a TBP stereochemistry (Fig. 11.16) is the one most likely to be empty, especially for the spin-paired d^7 and d^8 configurations. The metal d electrons are therefore mainly localized in the equatorial place; hence, the equatorial bonds may be expected to be longer than the apical. More up-to-date data on the pentahalides of

Table 11.20 Five-Coordinate Compounds

	Compound	Shape	Comments
Ti	$(Me_3N)_2TiBr_3$	TBP	Br's in trigonal plane
V	$VO(acac)_2$	SP	Short axial VO bond 156 pm
	$VO(SCN)_4$	SP	VO 162 pm
	$VO(1\ phenylbutane\text{-}1,3\text{-}dionato)$	SP	VO 161.2 pm
Nb, Ta	$NbCl_5$	TBP $\Big\}$	In vapor state only
	$TaCl_5$	TBP	
Mo	$MoCl_5$	TBP	In vapor state only
	$[Mo(acetate)_2]_2$	SP	Determined by Mo—Mo bond
Mn	$[Mn(CO)_5]^-$	TBP	From infrared
Re	$[ReOBr_4]^-$	SP	Due to ReO bond 180 pm
	$ReNCl_2(Ph_3P)_2$	Distorted SP	
	$[Re_2Cl_8]^{2-} \Big\}$	SP	Due to metal-metal bond
	$[Tc_2Cl_8]^{3-}$		
Fe	$Fe(CO)_5$	TBP	$Fe-C_{ax}$ 179 pm, $Fe-C_{eq}$ 179 pm
	$Fe(CO)_4(Ph_3P)$	TBP	Infrared evidence
	$Fe(terpyridyl)Cl_2$	Distorted SP	
	$[Fe(S_2CNEt_2)_2Cl]$	SP	FeS 232 pm, FeCl 227 pm (axial)
	$[Fe(S_2C_2(CN)_2)_2NO]$	SP	FeS 227 pm, FeN 156 pm (axial)
Ru	$RuCl_2(Ph_3P)_3$	SP	Pseudo-octahedral due to o hydrogen on one phenyl ring; RuP_{base} 223–237 pm, RuP_{ax} 200 pm
Co	$[Co(CH_3CN)_5]^+$	TBP	$Co-N_{ax}$ 184 pm, CoN_{eq} 188 pm
	$Co(CO)_3(Ph_3P)(C_2F_4H)$	TBP	
	$Co(Ph_3P)_3H(N_2)$	TBP	N_2 and H trans to each other
	$[Co(diphos)_2X]^+$	TBP	
	$[Co(OAsR_3)_4ClO_4]$	SP	
	$[Co(S_2C_6Cl_4)_2]_2$	SP	
	$CoBr_2(Ph_3P)_3$	TBP	
	$Co(paphy)Cl_2$	SP	
	$[Co(Ph_2CH_3AsO)_4ClO_4]ClO_4$	Distorted SP	
	$CoBr_3(R_3P)_2$	TBP	
	$[Co(MNT)_2L]$	SP	$L = C_5H_5N, R_3P, R_3As$
Rh	$Rh(Ph_3P)_3(CO)H$	TBP	
	$Rh(Ph_3P)_2MeI_2$	SP	

Table 11.20 (*cont.*)

	Compound	Shape	Comments
Ir	$[Ir(Ph_3P)_2COCl(SO_2)]$	SP	
	$[Ir(Ph_3P)_2COCl(O_2)]$	TBP	
Ni	$[Ni(CN)_5]^{3-}$	Both SP and TBP in same crystal	
	$NiBr_2(PMe_3)_3$	TBP	
	$NiBr_2$(triarsine)	SP	
	$[NiBrN(CH_2CH_2N(CH_3)_2)_3]$	TBP	
	$NiBr(diphos)(\pi\text{-}CH_2CMeCH_2]$	Distorted SP	
	$NiBr_3(Et_3P)_2$	TBP	
Pd	$Pd(PhMe_2P)_2I_2$	SP	Long axial bond
Pt	$[Pt(SnCl_3)_5]^{3-}$	TBP	
	$[Pt(QAS)I]^+$	TBP	(Also Ni, Pd)
Cu	$[Cu(bipy)_2I]I$	TBP	
	$[CuCl_5]^{3-}$	TBP	$CuCl_{ax}$ 230 pm, $CuCl_{eq}$ 239 pm
	$Cu(formate)_2$	SP	
	$Cu(dmg)_2$	SP	
	$[Cu(salicylaldehyde-ethylenedi-imine)_2]_2$	SP	
	$[Cu(NN'di\text{-}n\text{-}propyldi-thiocarbamate)_2]$	SP	
	$CuCl_2(H_2O)(o\text{-phen})$	TBP	
Ag	$Ag(C_6H_6)AlCl_4$	SP	

Mo, Ta, and Nb in the gaseous state would be interesting, because in these cases (d^1 and d^0) one may expect the apical bonds to be longer in the same way as for the nontransition elements.

In terms of simple electrostatic theory, a distorted SP (Fig. 11.15) is more stable than a regular shape. The distortion results in the metal atom being raised above the plane of the four coplanar ligands in the direction of the fifth ligand. For a $M-L$ bond length of around 200 pm, the expected distance of the metal ion above the plane of the four coplanar ligands is 48 pm based on a $L_{ap}ML_b$ angle of 104°. The fair agreement between the calculated and observed distortion (Table 11.21) may be fortuitous rather than an indication that the electrostatic model is an adequate description. For the SP stereochemistry, the apical and base bond lengths vary with respect to each other from compound to compound. For the VOL_4 complexes, the apical bond is shortest due to the VO multiple bonding, whereas for Cu(II) in square pyramid stereochemistry, it is generally longer, owing to the d-electron repulsion in the apical direction.

Table 11.21 Distortion in Square Pyramidal Complexes

Compound	Bond lengths in base (pm)	Distance of metal above square plane (pm)	
$NOFe[S_2C_2(CN)_2]_2$	FeS 227	50 above	S_4 plane
$(Ph_3P)_3RuCl_2$	RuP 237, 223	40 above	P_2Cl_2 plane
	RuCl 239		
$Co(paphy)Cl_2$	CoCl 228	40 above	N_3Cl plane
$[(Ph_2CH_3AsO)_4CoClO_4]^+ClO_4^-$	CoO 202	32 above	O_4 plane
$[Ir(Ph_3P)_2COCl(SO_2)]$	IrP 236	21 above	P_2CCl plane
	IrC 196		
	IrCl 237		
$Ni(Deas)_2$	NiN 204, 225	36 above	O_2N_2 plane
	NiO 196		
$Cu(NN'$-di-*n*-propyldithio-carbamate$)_2$	CuS 232	40 above	S_4 plane

The distortions in the SP and TBP structures often give shapes intermediate between both, and discussion as to which ideal shape the structure approaches is often rather pointless.

The *d* Configuration and Stereochemistry

Table 11.22 depicts the range of metal ions that form five-coordinate complexes. The d^8 configuration is clearly the most prolific in forming five-coordinate compounds. One reason is the readiness for d^8 square planar complexes to form 1:1 adducts with a fifth ligand, forming an SP structure. Another reason is supplied by the nine-orbital rule; if the eight electrons are spin paired, it leaves one *d*, the *s*, and three *p* orbitals for bonding. The ease with which these orbitals can be used is somewhat determined by the *d-p* energy separation. The separation is largest when the charge on the metal ion is greatest.[4] Hence, for the metal ions with the d^8 configuration,

Cr(-II)	Mn(-I)	Fe(O)	Co(I)	Ni(II)	Cu(III)
Mo(-II)	Tc(-I)	Ru(O)	Rh(I)	Pd(II)	
W(-II)	Re(-I)	Os(O)	Ir(I)	Pt(II)	Au(III)

the tendency toward five coordination increases to the left. Vertical trends are less easily predicted. The effective nuclear charge increases down a group, but this may not be the only influencing factor. The $d^9 \rightarrow d^8 p^1$ promotion energies (Table 11.23) indicate that for the cobalt and nickel series the tendency toward five coordination is first row > second row

Table 11.22 Five Coordination*

Group					Configuration						
	d^{10}	d^9	d^8	d^7	d^6	d^5	d^4	d^3	d^2	d^1	d^0
IV										Ti(III)	~~Ti(IV)~~
V									V(III)	V(IV)	~~V(V)~~ Nb(V) Ta(V)
VI			Cr(−II) Mo(−II)				Mo(II)		Cr(IV)	Mo(V) ~~W(V)~~	~~Cr(VI)~~ Mo(VI) W(VI)
VII			Mn(−I) Tc(−I)? Re(−I)?	Mn(O)		Mn(II) Re(II)?	Mn(III)? ~~Tc(2.5)~~ Re(III)		~~Re(V)~~		
VIII *a*		Fe(−I)	Fe(O) Ru(O) Os(O)	Fe(I)	~~Fe(II)?~~ Ru(II)	Fe(III)		Ru(V)	Os(VI)		Os(VIII)
b	Rh(−I)? Ir(−I)?	Co(O)?	~~Co(I)~~ ~~Rh(I)~~ ~~Ir(I)~~	~~Co(II)~~ Rh(II) Ir(II)?	Co(III)? Rh(III) Ir(III)						
c			~~Ni(II)~~ Pd(II) ~~Pt(II)~~	Ni(III)							
I	~~Ag(I)~~	~~Cu(II)~~	Au(III)								

* TBP represented by M(III); SP represented by M(II).

226

Table 11.23 $d^9 \rightarrow d^8 p$ Energies (eV)

Co(O)	Ni(I)	Cu(II)
0.85	2.90	6.01
Rh(O)	Pd(I)	Ag(II)
1.60	3.39	5.76
Ir(O)	Pt(I)	Au(II)
2.4	3.05	5.2

$\geqq$ third row, whereas for the copper series the tendency is third row > second row > first row. The electronegativity of the ligand may also affect the coordination number. The ligands with high electronegativity will increase the effective positive charge on the metal, which will reduce the tendency toward five coordination compared with four coordination.

The electron arrangements possible for the d^5 to d^8 configurations are given in Table 11.24 for both the TBP and SP shapes (based on the orbital

Table 11.24 Electron Configuration for SP and TBP Stereochemistries

	TBP				SP				
	(e'')	(e')	(a_1')	Σs_i	(e)	(b_2)	(a_1)	(b_1)	Σs_i
d^5 Weak	2	2	1	$\frac{5}{2}$	2	1	1	1	$\frac{5}{2}$
		or							
	3	2		$\frac{3}{2}$	3	1	1		$\frac{3}{2}$
Strong	4	1		$\frac{1}{2}$	4	1			$\frac{1}{2}$
d^6 Weak	3	2	1	2	3	1	1	1	2
Strong	4	2		1	4	2			0
						or			
					4	1	1		1
d^7 Weak	4	2	1	$\frac{3}{2}$	4	1	1	1	$\frac{3}{2}$
Strong	4	3		$\frac{1}{2}$	4	2	1		$\frac{1}{2}$
d^8 Weak	4	3	1	1	4	2	1	1	1
Strong	4	4		0	4	2	2		0

splitting in Table 11.19). Whether spinfree or spin-paired complexes form depends on the magnitude of the energy parameters Δ_2 and Δ for TBP and SP, respectively (Table 11.19). Some examples of both high- and low-spin compounds are given in Table 11.25. If the apical bond is weakened in a SP structure, the orbital splitting changes, as shown in Fig. 11.17, and Δ increases, with more chance of spin pairing occurring. In fact, this may be the situation for the SP form of $[Ni(CN)_5]^{3-}$, where the axial bond is 214 pm and the base bonds are 185 to 189 pm.

Table 11.25   Spin States of Five-Coordinate Complexes

Compound	Shape	d^n	S	Spinfree (s.f.) or spin paired (s.p.)
[FeCl(dtc)$_2$]	SP	d^5	$\frac{3}{2}$	s.f.
[Fe(Me$_6$tren)Br]$^+$	TBP	d^6	2	s.f.
[Fe(QAS)Br]$^+$	TBP	d^6	1	s.p.
[Fe(QP)NO$_3$]$^+$	TBP	d^6	1	s.p.
[Co(Me$_6$tren)Cl]$^+$	TBP	d^7	$\frac{3}{2}$	s.f.
[Co(Me-Sal)$_2$]$_2$	TBP	d^7	$\frac{3}{2}$	s.f.
[Co(QAS)Br]$^+$	TBP	d^7	$\frac{1}{2}$	s.p.
[Co(QP)NO$_3$]$^+$	TBP	d^7	$\frac{1}{2}$	s.p.
[Ni(Me$_6$tren)Br]$^+$	TBP	d^8	1	s.f.
[Ni(Et$_4$dien)Cl$_2$]	TBP	d^8	1	s.f.
[Ni(Salen-NEt$_2$)$_2$]	SP	d^8	1	s.f.
[Ni(CN)$_5$]$^{3-}$	TBP	d^8	0	s.p.
[Ni(CN)$_5$]$^{3-}$	SP	d^8	0	s.p.
[Ni(trias)Br$_2$]	SP	d^8	0	s.p.
[Ni(TDPME)X_2] (X = Cl, Br, CN)	?	d^8	0	s.p.

Magnetic and Spectral Properties

The use of magnetism and electronic spectra for the determination of the stereochemistry of five-coordinate compounds is not very well developed, and more information is necessary before stereochemical deductions can be made unambiguously. In the case of the TBP shape, the orbital splitting parameter ratio Δ_2/Δ_1 has been calculated as 1.44. However, a higher ratio is obtained experimentally (Table 11.26). Attempts to improve the theoretical figure by allowing for likely distortions actually give less satisfactory agreement. Hence, more study of the finer details of spectral properties and stereochemistry is required.

The energy of the $a_1 \rightarrow b_1$ transition for an SP complex increases with increase in the apical bond length, and may therefore be used as a measure of this type of distortion[8] (Fig. 11.17).

The TBP shape has been proposed for a number of carbonyl compounds on the basis of infrared measurements; examples are listed as follows: [Mn(CO)$_5$]$^-$, [NOMn(CO)$_4$]$^-$, Ph$_3M$Fe(CO)$_4$ (M = P, As, Sb), [HFe(CO)$_4$]$^-$, Fe(CO)$_3$(Ph$_3M$)$_2$ (M = P, As, Sb), Fe(CO)$_3$(CNR)$_2$, HCo(CO)$_4$, Co(CO)$_3$(Ph$_3$P)I, and Ru(CO)$_3$(Ph$_3$P)$_2$. The number of fundamental ν(C≡O) stretching modes can be determined readily by group theoretical methods (Table 11.27) and can be used as a guide to the molecular or ionic

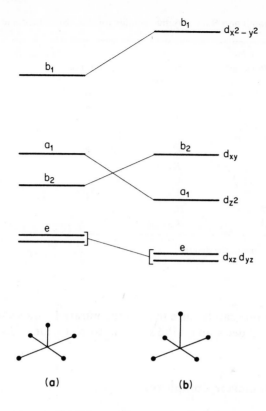

Fig. 11.17 Orbital splitting for: (a) square pyramidal structure, (b) distorted square pyramidal structure.

Table 11.26   Ligand Field Parameters for the TBP Stereochemistry

Compound	Δ_1	Δ_2	Δ_2/Δ_1	d^n
$[Fe(Me_6tren)Br]^+$	3,100	6,200	2	d^6 (s.f.)*
$Cu_4OCl_6(OPR_3)_4$	1,300	9,900	7.6	d^9
$CuCl_5^{3-}$	2,200	8,200	3.7	d^9
$[Cu(tren)OH]^+$	2,500	11,900	4.8	d^9
$[Cu(Me_6tren)X]^+$	$\sim 3,500$	$\sim 10,000$	2.9	d^9
$[Cr(Me_6tren)Br]Br$	$\sim 5,000$	$\sim 10,000$	2	d^4
$[Ni(TSP)Br]^+$	$\sim 5,600$	15,100	2.7	d^8 (s.p.)*
$[Ni(QAS)Br]^+$	$\sim 6,000$	15,800	2.6	d^8 (s.p.)
$[Pd(QAS)Br]^+$	7,600–10,800	19,000–22,600	2.1–2.5	d^8 (s.p.)

* s.f. = Spin free: s.p. = Spin paired.

Table 11.27   Infrared and Raman Active Modes for the TBP Stereochemistry

Compound	CO Position*	Symmetry	$\nu(C\equiv O)$ Infrared	$\nu(C\equiv O)$ Raman
$M(CO)_4L$	$(1, 2, 3, 5)$	C_{3v}	$2a_1 + e$	$2a_1 + e$
	$(1, 2, 4, 5)$	C_{2v}	$2a_1 + b_1 + b_2$	$2a_1 + b_1 + b_2$
$M(CO)_3L_2$	$(1, 2, 3)$	D_{3h}	e'	$a_1' + e'$
	$(1, 2, 5)$	C_s	$2a' + a''$	$2a' + a''$
	$(1, 4, 5)$	C_{2v}	$2a_1 + b_2$	$2a_1 + b_2$
$M(CO)_2L_3$	$(4, 5)$	D_{3h}	a_2''	a_1'
	$(1, 5)$	C_s	$2a'$	$2a'$
	$(1, 2)$	C_{2v}	$a_1 + b_2$	$a_1 + b_2$
$M(CO)L_4$	5	C_{3v}	a_1	a_1
	1	C_{2v}	a_1	a_1

* CO positions according to the diagram .

shape, provided one can be sure that all the infrared bands can be resolved. However, the final decision on the shape must rest with an X-ray structural analysis.

Dinuclear Complexes

The dinuclear complexes, such as $[M_2X_8]^{n-}$ and $[M_2(RCOO)_4]^{n+}$ (X is a halogen), have the structures given in Figs. 11.18(a) and (b). The environment of the metal ion, owing to the metal-metal bond, is a distorted square pyramid. The acetates of divalent Cu and Cr have a similar structure to that shown in Fig. 11.18(b), except that the metal atom achieves sixfold coordination by coordinating water molecules at extremities of the dimer. True five coordination exists in the compounds $[Mo_2(acetate)_4]$, $[Mo_2Cl_8]^{4-}$, $[Re_2X_8]^{2-}$ ($X = Cl, Br$), and $[Tc_2Cl_8]^{3-}$. The five coordination is maintained in $[Re_2X_8]^{2-}$ by the sixth position being blocked by the halide ligands. The structure of the halogeno- complexes is such that the two MX_4 groups are in the eclipsed position, suggesting a δ interaction between the metal atoms similar to that postulated for $Cu_2(acetate)_4 2H_2O$. Destruction of the δ bond in the complex $Re_2Cl_5(2,5\text{-dithiahexane})_2$ gives rise to the more favorable staggered arrangement of the ReX_4 and ReS_4 groups, with respect to each other.

A number of dimeric copper complexes achieve an SP structure (cf. Table 11.20) by forming the structural skeleton given in Fig. 11.19. The apical bond is generally 40 to 60 pm longer than the four planar bonds.

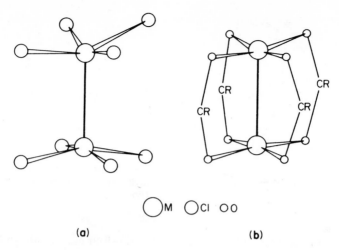

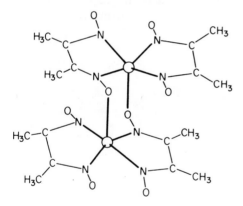

Fig. 11.18 Structure of (a) the $[Re_2Cl_8]^{2-}$ anion, and (b) $[M_2(RCO_2)_4]$.

Fig. 11.19 Structure of $[Cu(dimethylglyoxime)_2]_2$.

Steric Effects

Both square planar and square pyramidal complexes can result from the blocking of axial coordination sites by bulky ligands. A good example of this occurs in the square pyramidal complex $RuCl_2(Ph_3P)_3$. A phenyl group, of one of the phosphines, blocks the sixth position, although it is debatable whether an ortho hydrogen is bonded in the position because The Ru···H distance is 256 pm, which is the same as the sum of the van der Waals' radii (Fig. 11.20). The complex $PdCl(cpmH)(cpm)$ is an example in

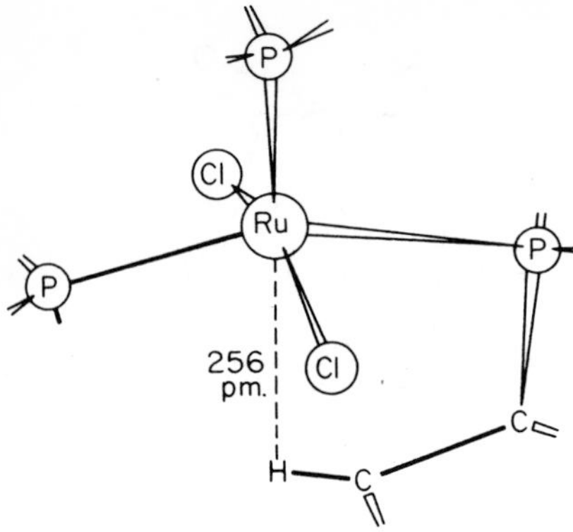

Fig. 11.20   Position of an α-carbon hydrogen in the structure of $RuCl_2(Ph_3P)_3$.

which it is difficult to decide whether the metal is four coordinate and square planar or pseudo-five coordinate and square pyramidal (Fig. 11.21). The protonated nitrogen atom is in a position where the proton may be bonded to the palladium.

The interaction of ligand protons with metal ions is not uncommon and needs to be taken into account when we consider the shape of molecules that appear to have vacant coordination sites around the metal.

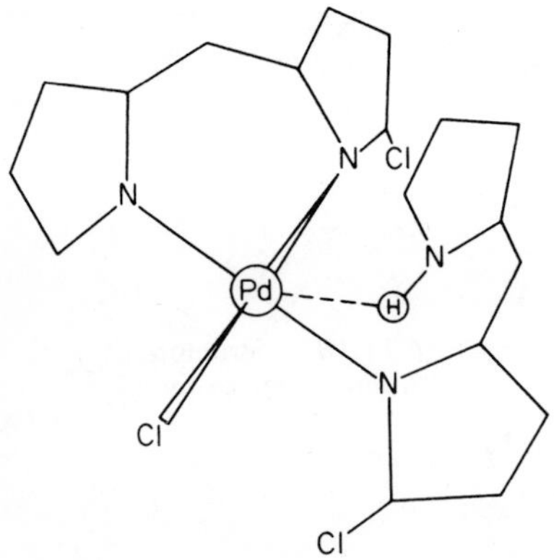

Fig. 11.21   Stereochemistry of Pd in the complex PdCl(cpmH)(cpm).

Reaction Mechanisms[3]

The increasing number of examples of stable five-coordinate complexes provides some justification for postulating five-coordinate intermediates in ligand replacement reactions. The dissociative S_N1 mechanism proposed for ligand substitution in octahedral complexes is

$$MA_6 \longrightarrow MA_5 + A \xrightarrow{B} MA_5B$$

five-coordinate
intermediate

For example, the reaction

$$cis[\text{Co(en)}_2\text{OHCl}]^+ + \text{H}_2\text{O} \rightarrow cis[\text{Co(en)}_2\text{OHH}_2\text{O}]^{2+} + \text{Cl}^-$$

gives 100% cis product, and the proposed mechanism is

$$\text{HO}-\text{Co} \xrightarrow[\text{Cl}]{} \xrightarrow{-\text{Cl}^-} \text{HO}-\text{Co} \xrightarrow{\text{H}_2\text{O}} \text{HO}-\text{Co}$$

cis   square pyramid   cis

A trigonal bipyramid intermediate would lead to a mixture of cis and trans products. A similar reaction for $trans[\text{Co(en)}_2\text{OHCl}]^+$ gives 75% *cis*- and 25% $trans[\text{Co(en)}_2\text{OH(H}_2\text{O)}]^{2+}$, which suggests a **TBP** intermediate

$$\text{HO}-\text{Co}-\text{Cl} \xrightarrow{-\text{Cl}} \text{HO}-\text{Co} \xrightarrow{+\text{H}_2\text{O}} \text{cis and trans products}$$

trans   cis

trigonal bipyramid

The water can attack anywhere around the triangular plane, and the ratio of cis to trans is theoretically $2:1$. If the intermediate is a SP, only one site is open to attack, viz., the base of the square pyramid, and therefore only one product would be formed. Although these idealized schemes are probably oversimplified, there appear to be good reasons for postulating five-coordinate intermediates.

Also, substitution reactions for square planar complexes probably proceed by a S_N2 associative mechanism, where a five-coordinate intermediate species is formed.

11.5 SIX COORDINATION

Six coordination is by far the most common coordination number for the transition metals (cf. Table 11.28). The coordination number occurs for all the elements and over a wide range of oxidation states and d electron configurations. The predominant octahedral geometry has cubic symmetry and

Table 11.28 Six Coordination

Group		Configuration									
	d^{10}	d^9	d^8	d^7	d^6	d^5	d^4	d^3	d^2	d^1	d^0
IV						Ti(−I)	Ti(O)		Ti(II)	Ti(III)	Ti(IV)
							Zr(O)			Zr(III)	Zr(IV)
											Hf(IV)
V					V(−I)	V(O)	V(I)	V(II)	V(III)	V(IV)	V(V)*
					Nb(−I)					Nb(IV)	Nb(V)
					Ta(−I)					Ta(IV)	Ta(V)
VI				Cr(−I)	Cr(O)	Cr(I)	Cr(II)	Cr(III)	Cr(IV)	Cr(V)	Cr(VI)*
					Mo(O)	Mo(I)	Mo(II)	Mo(III)	Mo(IV)*	Mo(V)	Mo(VI)*
					W(O)		W(II)?	W(III)	W(IV)*	W(V)	W(VI)*
VII		Mn(−II)?	Mn(−I)?	Mn(O)	Mn(I)	Mn(II)	Mn(III)	Mn(IV)			
				Tc(O)	Tc(I)	Tc(II)	Tc(III)	Tc(IV)			
				Re(O)	Re(I)	Re(II)	Re(III)	Re(IV)	Re(V)	Re(VI)*	Re(VII)

	(O)	(I)	(II)	(III)	(IV)	(V)	(VI)	(VII)	(VIII)
VIII									
a	Fe(O)	Fe(I)	Fe(II)	Fe(III)	Fe(IV)				
			Ru(II)	Ru(III)	Ru(IV)	Ru(V)	Ru(VI)		
	Os(O)		Os(II)	Os(III)	Os(IV)	Os(V)	Os(VI)	Os(VII)	Os(VIII)
b		Co(I)	Co(II)	Co(III)	Co(IV)				
			Rh(II)	Rh(III)	Rh(IV)				
		Ir(II)	Ir(III)	Ir(IV)	Ir(V)	Ir(VI)			
c			Ni(II)	Ni(III)	Ni(IV)				
			Pd(II)		Pd(IV)				
			Pt(II)		Pt(IV)	Pt(V)	Pt(VI)		
I		Cu(I)	Cu(II)	Cu(III)					
		Ag(I)		Ag(III)					
				Au(III)					

* Examples of trigonal prism stereochemistry.

—— Main examples of octahedral stereochemistry.

- - - - Structures known.

is an ideal arrangement for placing six ligands around a metal ion in order to minimize steric interactions. This, together with a favorable ligand field stabilization energy and good orbital overlap in the M—L bonds, contributes to the wide occurrence of six coordination and octahedral stereochemistry. It is especially evident for the metal ions of the second and third transition series, where the large metal radius is an important factor.

Octahedral Stereochemistry

It is relatively easy to obtain evidence for octahedral coordination, because the high cubic symmetry lends itself to extensive theoretical study. Therefore, the interpretation of spectral and magnetic data can often be reliably used in structural studies.

Configurations d^4 to d^7 inclusive occur in both the high- and low-spin states for the first row transition metals, whereas spin pairing invariably occurs for the second and third row transition metals. Metal ions with half-filled or entirely filled subshells strongly favor the octahedral coordination (Table 11.29). Electron correlation assists in stabilizing the complex in these cases.

Table 11.29   Metal Ions Favoring Octahedral
                    Coordination

Configuration	Example
t_{2g}^3	V(II), Cr(III), Mn(IV)
	Mo(III), Tc(IV)
	W(III), Re(IV)
$t_{2g}^3 e_g^2$	Mn(II), Fe(III)
	(Only for first row transition metals)
t_{2g}^6	Fe(II), Co(III), Ni(IV)
	Ru(II), Rh(III), Pd(IV)
	Os(II), Ir(III), Pt(IV)
$t_{2g}^6 e_g^2$	Ni(II)
	Pd(II)
	Pt(II)

Distortion of Octahedron. A number of effects can distort an octahedron. The influence of an unsymmetrical d-electron arrangement on an octahedral stereochemistry has already been discussed in Section 7.4 (often called Jahn-Teller effect). The configurations where the distortion is greatest are $t_{2g}^3 e_g^1$ [Cr(II), Mn(III)], $t_{2g}^6 e_g^1$ [Co(II)], and $t_{2g}^6 e_g^3$ [Cu(II)]. Some examples of

Table 11.30   Some Distorted Octahedral Compounds

Compound	Bond lengths (pm)
$CrCl_2$	4CrCl 240, 2CrCl 291
MnF_3	2MnF 179, 2MnF 191, 2MnF 209
$CuCl_2$	4CuCl 230, 2CuCl 295
$K_2CuCl_4 2H_2O$	2CuOH$_2$ 201, 2CuCl 229, 2CuCl 293
$Cu(NH_3)_2Cl_2$	2CuN 195, 4CuCl 276

complexes with the distorted environment are given in Table 11.30. The distortion, in most cases, can be considered as tetragonal, where two trans bonds are longer or shorter than the other four bonds. The point group symmetry is D_{4h}. For mixed-ligand complexes, the stronger ligands bond at the points of least distortion, as, for example, in $K_2CuCl_4·2H_2O$ (Fig. 11.22).

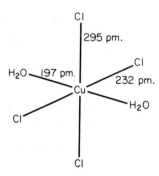

Fig. 11.22   Structure of the octahedral anion $[CuCl_4 2H_2O]^{2-}$ in $K_2CuCl_4 2H_2O$.

The other main octahedral distortion is trigonal, corresponding to an extension, or compression, along the C_3 axis. The resulting stereochemistry is a trigonal antiprism, with point group symmetry D_{3d} or C_{3v}. This stereochemistry occurs for a 1, 2, 3 MA_3B_3 complex (also called cis or facial) (Fig. 11.23(a)). An example is 1, 2, 3 $IrCl_3(Et_2S)_3$. The alternative arrangement, 1, 2, 6 MA_3B_3 (trans or meridonal) corresponds to a rhombohedral distortion of the octahedron [Fig. 11.23(b)].

Electronic spectra can be used to distinguish octahedral complexes having these different types of distortion. For example, for Co(III) (t_{2g}^6), the two lowest energy spin-allowed transitions in a regular octahedral environment are $^1A_{1g} \rightarrow {}^1T_{1g}$ and $^1A_{1g} \rightarrow {}^1T_{2g}$. The T_{1g} and T_{2g} terms split in tetragonal and rhombohedral fields, as Fig. 11.24 shows. Therefore, 2 and 6 absorptions are expected for the 1, 2, 3 CoA_3B_3 and 1, 2, 6 CoA_3B_3 compounds, respectively. Owing to problems of resolution, it is generally found that the 1, 2, 6 isomers have broader and less symmetric bands than the 1, 2, 3 isomers.

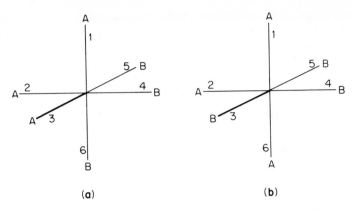

Fig. 11.23 Stereochemistry of MA_3B_3 complexes: (a) 1,2,3, or *cis* or *fac*; (b) 1,2,6, or *trans* or *mer*.

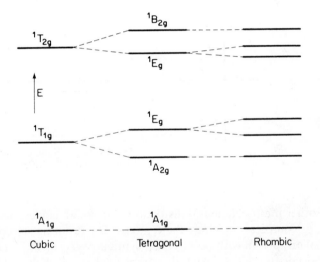

Fig. 11.24 Splitting of terms for the d^6 spin paired configuration for distorted octahedral structures.

The α and β forms of Co(glycinate)$_3$ have been distinguished on this basis. The spectral differences for cis and trans CoA$_4$B$_2$ complexes have been discussed in Chapter 7. In electronic spectral studies of isomeric compounds, we should remember that band broadening and splitting may occur for other reasons, such as spin-orbit coupling and molecular vibrations.

Infrared Measurements. Vibrational spectroscopy has been employed to investigate regular octahedral structures. Since the number of infrared and

Raman fundamental active modes are determined by the symmetry of a molecule, it is possible, in favorable cases, to compare the number of fundamental modes predicted and observed. Such an analysis is not restricted to an octahedral complex, but in view of its high symmetry, it is a particularly favorable case (Table 11.31). It is normally necessary to use both infrared and Raman data to obtain definitive results.

Catalysis. A number of transition metal homogeneous catalysts involved in chemical reactions depend on the stereochemistry of the metal ion. Pseudovacant sites in octahedral complexes are potential reaction sites. This is illustrated by one of the mechanisms proposed for producing polyethylene, using Ziegler-Natta catalysts.

The alternating vacant site is a necessary requirement of the mechanism. In the same way, the vacant or solvent site in the complex $Rh(Ph_3P)_3Cl$ is apparently important in the mechanism proposed for the catalytic hydrogenation of ethylene.

Table 11.31   Infrared (ir) and Raman (R) Active Modes for Octahedral Compounds

Activity	R	R	ir	ir	R	Inactive
Mode*	v_1	v_2	v_3	v_4	v_5	v_6
Symmetry	A_{1g}	E_g	T_{1u}	T_{1u}	T_{2g}	T_{2u}
MoF$_6$	741	643	741	264	306	(190)†
WF$_6$	769	670	712	256	322	(216)†
PtF$_6$	655	601	705	273	242	(211)†
[PtCl$_6$]$^{2-}$	344	320	343	182	162	
[ReCl$_6$]$^{2-}$	346	(275)†	313	172	159	

* See diagrams below.
† Calculated values.

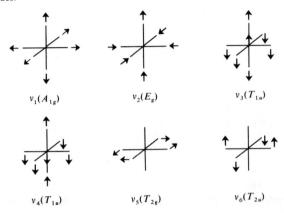

$$v_1(A_{1g}) \qquad v_2(E_g) \qquad v_3(T_{1u})$$

$$v_4(T_{1u}) \qquad v_5(T_{2g}) \qquad v_6(T_{2u})$$

Metal Halides. A glance at *Interatomic Distances*, compiled by Sutton,[9,10] will indicate the great number of octahedral complexes whose structures have been determined. One class of compounds, the binary metal halides, irrespective of stoichiometry, has a tendency to form structures with the metal ion in sixfold coordination and octahedral stereochemistry in the solid state (Table 11.32). The method by which the metal halides achieve six coordination are indicated in the table and in Figs. 11.25(a) through (i). The dominating influence of six coordination is quite apparent. Some of the halides of the heavy transition metals also have metal-metal interactions. This often leads to coordination numbers greater than six, as, for example, found for the subhalides of niobium and tantalum, dihalides of molybdenum and tungsten, and trihalides of rhenium.

Trigonal Prismatic Stereochemistry

The trigonal prism (Fig. 11.26) is predicted to be less stable than the octahedron (skewed trigonal prism) because of the closer contacts between the six donor atoms. The shape occurs for the disulfides of

Table 11.32 Transition Metal Halides with Octahedral Structures

---

d^0 $(NbF_5)_4$,[1] $(TaF_5)_4$,[1] $NbCl_5$,[2] $TaCl_5$,[2] MoF_6, WF_6

d^1 TiF_3,[3] $TiCl_3$,[4] $TiBr_3$,[4] $\beta ZrCl_3$,[5] NbI_4,[6] TaI_4,[6] $(MoF_5)_4$,[1] $MoCl_5$,[2] WCl_5[2]

d^2 $TiCl_2$,[7] TiI_2,[7] VF_3,[3] VCl_3,[4] NbF_3,[8] TaF_3,[8] OsF_6

d^3 VCl_2,[7] CrF_3,[3] $CrCl_3$,[9] $CrBr_3$,[4] CrI_3,[9] MoF_3,[8] $MoCl_3$,[6] $MoBr_3$,[6] $TcCl_4$,[10] β-$ReCl_4$,[11] $(RuF_5)_4$[1]

d^4 CrF_2,[12] $CrCl_2$,[12] $CrBr_2$,[12] CrI_2,[12] Cr_2F_5,[12] MnF_3[12]

d^5 MnF_2,[13] $MnCl_2$,[14] MnI_2,[7] FeF_3,[3] $FeCl_3$,[4] $FeBr_3$,[4] RuF_3[15]

d^6 FeF_2,[13] $FeCl_2$,[14] $FeBr_2$,[7] FeI_2,[7] CoF_3,[3] RhF_3,[15] $RhCl_3$, IrF_3[15]

d^7 CoF_2,[13] $CoCl_2$,[14] $CoBr_2$,[7] CoI_2[7]

d^8 NiF_2,[13] $NiCl_2$,[14] $NiBr_2$,[7] NiI_2,[14] PdF_2,[13] $PtCl_2$

d^9 CuF_2,[12] $CuCl_2$,[12] $CuBr_2$[12]

---

[1] See Fig. 11.25(a).

[2] Dimers; see Fig. 11.25(b).

[3] Intermediate h.c.p. and ReO_3 structure.

[4] BiI_3 structure; Fig. 11.25(c).

[5] Distorted h.c.p.

[6] Metal-metal interaction; Fig. 11.25(d).

[7] CdI_2 structure; Fig. 11.25(e).

[8] Cubic ReO_3 structure; Fig. 11.25(f).

[9] $CrCl_3$ structure; Fig. 11.25(c).

[10] See Fig. 11.25(g).

[11] $W_2Cl_9^{3-}$-type structure; Fig. 10.5.

[12] Distorted octahedral environment, rutile, Fig. 11.25(h).

[13] Distorted octahedral environment, rutile, Fig. 11.25(i).

[14] $CdCl_2$ structure; Fig. 11.25(e).

[15] H.c.p. structure.

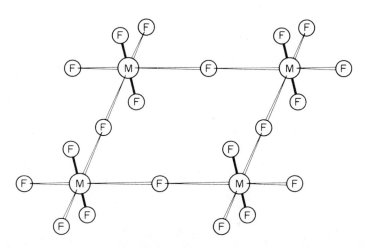

Fig. 11.25(a) Structure of $(MF_5)_4$.

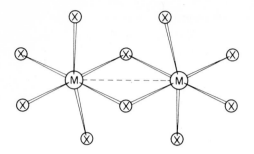

Fig. 11.25(b)   Structure of $(MX_5)_2$ halides.

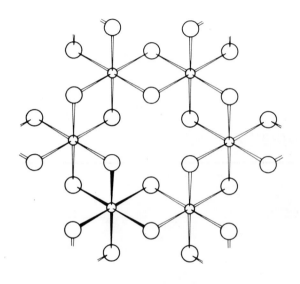

○ M   ○ X

Fig. 11.25(c)   Structure of MX_3 halides.

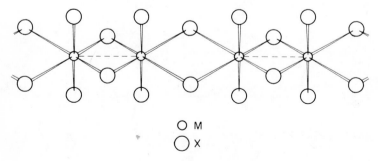

○ M
○ X

Fig. 11.25(d)   Structure of MX_4 halides.

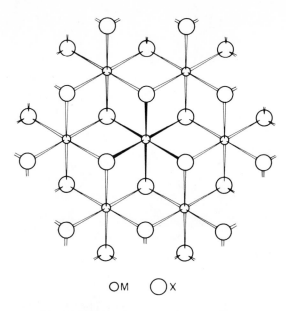

O M   ◯ X

Fig. 11.25(e)   Structure of *MX₂* halides.

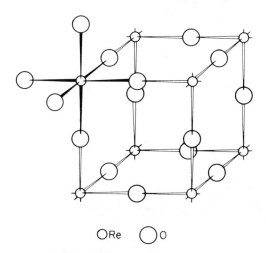

O Re   ◯ O

Fig. 11.25(f)   ReO₃ structure.

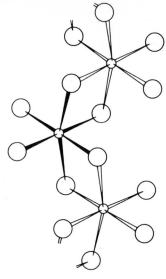

○ Tc   ◯ Cl   **Fig. 11.25(g)**   Structure of $TcCl_4$.

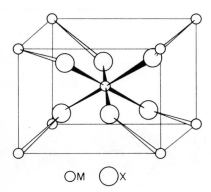

○ M   ◯ X   **Fig. 11.25(h)**   Distorted rutile structure.

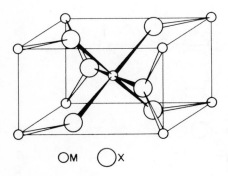

○ M   ◯ X   **Fig. 11.25(i)**   Distorted rutile structure.

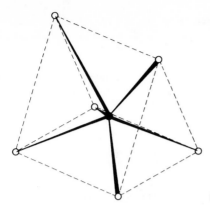

Fig. 11.26 Trigonal prism.

molybdenum and tungsten, and may be a consequence of crystal packing. If for some reason the ligand donor atoms need to be close together, as may be the case for a bidentate ligand, then a trigonal prism may be formed. This is the case for the dithiolato- ligands

where $R = $ CN (MNT), Ph (SDT), CF_3, H, CH_3, and

where $R' = $ H (BDT) and CH_3 (TDT). These ligands form a number of tris- complexes of the type $[M(S-S)_3]^{0 \text{ or } -}$, with metal ions such as Re, Mo, W, V, and Cr, which all have the trigonal prismatic structure (Table 11.33).

For the cases where the structures are known, a short $S\cdots S$ distance of 305 to 311 pm occurs between the sulfur atoms of the same ligand and sulfur atoms of adjacent ligands, suggesting a partial S—S bond. This is probably the main reason for the shape, as only in a trigonal prism can the sulfur atoms get close enough for bonding. Another factor may be the considerable M-ligand π bonding possible in the trigonal prismatic structure.

The M—S distances are relatively constant from complex to complex: V—S, 233.8 pm in V(SDT)$_3$; Mo—S, 233 pm in Mo(S$_2$C$_2$H$_2$)$_3$; and Re—S, 232.5 pm in Re(SDT)$_3$.

Table 11.33   Trigonal Prismatic Structures

Compound	M—S (pm)	S—S (pm) Intramolecular	S—S (pm) Intermolecular	$\widehat{SMS}$
Re[S$_2$C$_2$(C$_6$H$_5$)$_2$]$_3$	232.5	303.3	305.0	81.4°
Mo(S$_2$C$_2$H$_2$)$_3$	233	310	311	82.5°
V[S$_2$C$_2$(C$_6$H$_5$)$_2$]$_3$	233.8	306.0	306.5	81.6°
(Distorted trigonal prism)				
V[S$_2$C$_2$(CN)$_2$]$_3^{2-}$[(CH$_3$)$_4$N$^+$]	236	313	320	
Intermediate trigonal prism and octahedron				
Fe[S$_2$C$_2$(CN)$_2$]$_3^{2-}$	226.1	314.7	319	88.0°
Octahedral				

A simplified MO energy level diagram for Re(SDT)$_3$ is given in Fig. 11.27.[11] The ground state $(4e')^4(2a_2')^2(3a_1')^1$ agrees with the magnetic and ESR data. The complex (as for others) can be reduced to give the species

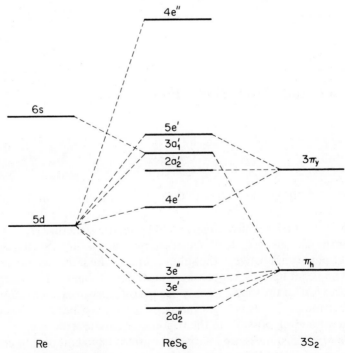

Fig. 11.27   Simplified molecular orbital energy diagram for Re(SDT)$_3$.

$Re(SDT)^{n-}$, where $n = 1, 2,$ and 3. The second and third additional electrons will enter the strongly antibonding $5e'$ orbital, which will produce strong $S\cdots S$ repulsions and may cause the rupture of the S—S bond. Therefore, it is predicted that the stereochemistry of the reduced complexes will revert to octahedral. This is found to be so, and $V(MNT)_3^{2-}$ has a structure intermediate between a trigonal prism and octahedron, whereas $Fe(MNT)_3^{2-}$ is octahedral.

11.6   SEVEN COORDINATION

Seven coordination is not common and is mainly limited to a few metals on the left of the transition metal series, in particular, the heavier members (Table 11.34). The three possible shapes (Section 7.4) are the 1,3,3 structure

Table 11.34   Seven Coordination

| Group | Configuration | | | | | |
	d^5	d^4	d^3	d^2	d^1	d^0
IV						Ti(IV)
						Zr(IV)
						Hf(IV)
V	V(O)					
						Nb(V)
						Ta(V)
VI		Cr(II)		Cr(IV)		
		Mo(II)	Mo(III)			Mo(VI)
		W(II)	W(III)			W(VI)
VII	Mn(II)					
		Re(III)		Re(V)	Re(VI)	Re(VII)
VIII	Fe(III)					

based on an octahedron with the seventh bond protruding from a face; the 1,4,2 structure, which is based on a trigonal prism with the seventh bond protruding from one of the square faces; and the 1,5,1 structure, which is a pentagonal bipyramid (see Fig. 4.6). Table 11.35 lists a number of seven-coordinate compounds classified according to their shape. The most common shape is the pentagonal bipyramid, but it is distorted in all cases. The complexes listed in Table 11.36 are probably seven coordinate.

Seven coordination occurs mainly for elements with d-electron configurations in the range of from d^0 to d^4. In these cases, sufficient bonding

Table 11.35   Some Seven-Coordinate Compounds

Shape	Electron configuration	Metal ion	Example
1, 3, 3, Octahedron	d^0	Zr(IV)	ZrO_2 ⎫ very distorted
$+1$	d^0	Hf(IV)	HfO_2 ⎭
(d^5sp, d^3sp^3)	d^0	Nb(IV)	$[NbOF_6]^{3-}$
1, 4, 2 Trigonal	d^0	Zr(IV)	$[Zr_2F_{13}]^{5-}$
prism $+1$	d^0	Nb(V)	$[NbF_7]^{2-}$
$(d^4sp^2, d^4p^3, d^5p^2)$	d^0	Ta(V)	$[TaF_7]^{2-}$
	d^5	Mn(II)	$[Mn(H_2O)(EDTA)]^-$
1, 5, 1 Pentagonal	d^0	Zr(IV)	$[ZrF_7]^{3-}$
bipyramid	d^2	Cr(IV)	$Cr(O_2)_2(NH_3)_3$
(d^3sp^3)	d^0	W(VI)	$K_2[W_2(O_2)_4(H_2O)_2O]$
	d^4	Re(III)	$Re_3X_{12}^{3-}$ ($X = $ Cl, Br)
	d^4	Re(III)	$Re_3X_{11}^{2-}$ ($X = $ Cl, Br)
	d^4	Re(III)	Re_3X_9 ($X = $ Cl, Br, I)
	d^4	Re(III)	$Re_3Cl_9L_3$
	d^5	Fe(III)	$[Fe(H_2O)(EDTA)]^-$
	d^5	Fe(III)	*Trans*[Fe 1,2-diaminocyclohexane N,N'tetraacetate,aquo]$^-$

orbitals are available, but spin pairing appears to be necessary for the d^3 and d^4 configurations. Therefore, they are mainly restricted to the second and third row transition metals. Some seven-coordinate complexes in which the metal ion has the d^5 spinfree configuration are known. The reason for their existence is probably related to a zero crystal field stabilization energy for the d^5 spinfree configuration. In this situation, the main barrier to seven coordination over six is that owing to steric factors. Provided these are not dominating, seven coordination may be possible. The expected shape would be the most symmetrical—namely, a pentagonal bipyramid structure.

The ligand most commonly associated with seven coordination is the fluoride, presumably because of its small size and ability to stabilize high oxidation states where the number of d electrons is small. The peroxide ion $(O_2)^{2-}$ can be considered a bidentate ligand from the point of view of coordination number. Since the oxygens are bonded in pairs, the space occupied is less than two separate oxygen atoms, which probably accounts for the fact that O_2^{2-} is associated with high coordination numbers in transition metal complexes.

Table 11.36   Possible Seven-Coordinate Compounds

Compound	Metal ion	Electron configuration
$[Ti(O_2)F_5]^{3-}$	Ti(IV)	d^0
$V(CO)_6(Ph_3P)$	V(O)	d^5
$[Cr(CO)_2(diars)_2X]X$	Cr(II)	d^4
$[Mo(CO)_2(diars)_2X]X$ X = halogen	Mo(II)	d^4
$[W(CO)_4(diars)X]X$	W(II)	d^4
$[W(CO)_3(diars)Br_2]$	W(II)	d^4
$K_4[Mo(CN)_7]2H_2O$	Mo(III)	d^3
$[MoF_7]^-$	Mo(VI)	d^0
$[WF_7]^-$	W(VI)	d^0
$[Re(CO)(diars)_2X_2]^-$	Re(III)	d^4
$ReOCl_3TAS$	Re(V)	d^2
$[ReOCl_6]^{2-}$	Re(VI)	d^1
ReF_7	Re(VII)	d^0

Rhenium(III) Cluster Compounds

The trinuclear rhenium(III) complexes Re_3X_9, $Re_3X_{12}^{3-}$, $Re_3X_{11}^{2-}$ (Table 11.35) contain seven coordinate rhenium atoms, each metal bonding to five halogens and two rhenium atoms in approximately a pentagonal bipyramid (Fig. 11.28). The bonding around each rhenium may be considered as 7σ bonds and 2π bonds (one to each of the two neighboring rhenium atoms). In this way, rhenium achieves maximum use of its bonding

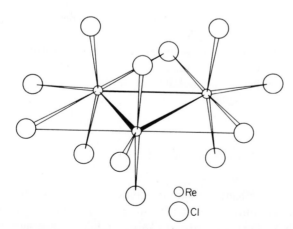

Fig. 11.28   Structure of the $[Re_3Cl_{12}]^{3-}$ anion.

orbitals and electrons to give seven coordination. The structure is under steric strain because the terminal in-plane ligands are at a longer distance from the metal than the rest and can be readily replaced or removed entirely to form the species $Re_3X_{11}^{2-}$ and $Re_3Br_{10}^-$, where some of the rhenium ions are now six coordinate.

Ligand Substitution

Ligand substitution in octahedral complexes has already been discussed in terms of an S_N1 reaction (p. 232) and five coordinate intermediates. Examples of the S_N2 mechanism are also known, where it is predicted that the intermediates are seven-coordinate compounds, either an octahedral wedge or a pentagonal bipyramid. For example, the base hydrolysis of [Co(II)EDTA]$^-$ appears to involve an S_N2 mechanism, as does the racemization of the optically active complex. The mechanism of the racemization is thought to involve a water molecule:

7-coordinate
intermediate

From a consideration of relative CFSE's for an octahedron and the two seven-coordinate shapes (Table 11.37), we may predict that the following d-electron configurations may undergo ligand substitution by an S_N2 mechanism: octahedron $\rightarrow$ pentagonal bipyramid d^0, d^1, d^2, d^5, d^6, and d^7, all spinfree; octahedron $\rightarrow$ octahedral wedge d^0, d^1, d^2, d^4 spinfree and paired, d^6 spinfree, d^7 spinfree and paired, and d^9.

11.7   EIGHT COORDINATION[12,13]

Eight-coordinate transition metal complexes are not as uncommon as one might expect. A number of structural investigations of eight-coordinate compounds have been carried out (Table 11.38). In addition, the chemical and physical properties of a number of compounds indicate eight

Table 11.37   Crystal Field Stabilization Energies

Configuration	Octahedron	Pentagonal bipyramid	Octahedral wedge
		Weak Field	
d^0	0	0	0
d^1	4	5.28	6.08
d^2	8	10.56	8.68
d^3	12	7.74	10.20
d^4	6	4.93	8.79
d^5	0	0	0
d^6	4	5.28	6.08
d^7	8	10.56	8.68
d^8	12	7.74	10.20
d^9	6	4.93	8.79
d^{10}	0	0	0
		Strong Field	
d^4	16	13.02	16.26
d^5	20	18.30	18.86
d^6	24	15.48	20.37
d^7	18	12.66	18.94

Table 11.38   Some Eight-Coordinate Compounds

Compound	Structure	Comments
$Ti(NO_3)_4$	Dodecahedron	Ti—O, 210 pm
$[TiCl_4(diars)_2]$	Dodecahedron	θ_a 36.4°, θ_b 72.5°
$[MX_4(diars)_2]$		
(M = Ti, Zr, Hf, V, Nb; X = Cl, Br)	Dodecahedron	Isomorphous to Ti/Cl compound
$Zr(acac)_4$	Antiprism	ZrO, 219.8 pm; $\theta = 57.3°$
$[Zr(C_2O_4)_4]^{4-}$	Dodecahedron	ZrO, 223, 216.8 pm; $\theta_a = 35.2°$;
Hf Complex isomorphous		θ_b, 73.5°
$[Zr(NTA)_2]^{2-}$	Dodecahedron	ZrN, 243.9 pm; ZrO 225, 212.8 pm
Hf Complex isomorphous		
ZrF_4	Antiprism	ZrF, 211 pm polymeric
		Hf Compound isomorphous
Zr, $HfOX_28H_2O$ (X = Cl, Br)	Antiprism	Polymeric
HfF_43H_2O	Antiprism	Polymeric
$Zr(IO_3)_4$	Antiprism	ZrO, 220.6 pm polymeric
$Zr(SO_4)_24H_2O$	Antiprism	ZrO, 218 pm polymeric

Table 11.38 *(cont.)*

Compound	Structure	Comments
$Zr(OH)_2SO_4$	Antiprism	ZrO, 219 pm polymeric
K_2ZrF_6	Dodecahedron	ZrF, 218 pm; $\theta_a = 35.3°$, $\theta_b = 74.6°$
ZrO_2	Dodecahedron	ZrO, 245.5, 206.5 pm polymeric
$[ZrF_4(H_2O)_3]_2$	Dodecahedron	ZrF, 198–222 pm; ZrO, 217–225 pm
$ZrSiO_4$	Dodecahedron	ZrO, 215, 229 pm polymeric
$Zr_2(OH)_2(SO_4)_3(H_2O)_4$	Dodecahedron	ZrO, 222, 219 pm
$Zr(OH)_2(NO_3)_2(H_2O)_4$	Dodecahedron	ZrO, 222 pm polymeric
$[ZrF_8]^{4-}$	Dodecahedron	ZrF, 216, 205 pm; $\theta_a = 43.0°$; $\theta_b = 65.6°$
$[M(O_2)_4]^{3-}$ ($M =$ V, Nb, Ta)	Dodecahedron	Isomorphous with Cr compound
$[TaF_8]^{3-}$	Antiprism	TaF, 198 pm
Ta, Nb_6X_{14} structures		
$[Cr(O_2)_4]^{3-}$	Dodecahedron	CrO, 187, 197 pm; $\theta_a = 43.4°$; $\theta_b = 86.8°$
$[W(CN)_8]^{4-}$	Antiprism	
$[W(CN)_8]^{3-}$	Dodecahedron	MoC, 215 pm; $\theta_a = 36.0°$; $\theta_b = 71.8°$
$[Mo(CN)_8]^{3-}$	Dodecahedron	Isomorphous to W compound
ReF_8^{3-}	Antiprism	ReF, 190 pm
$Ph_4As[Co(NO_3)_4]$	Dodecahedron	CoO, 207(4), 245 pm(4); $\theta_a = 47°$; $\theta_b = 81°$
$Ph_4As[Co(CF_3CO_2)_4]$	Dodecahedron	CoO, 200(4), 311 pm(4); $\theta_a = 48.5°$; $\theta_b = 81.5°$
$Ca[Cu(CH_3CO_2)_4]$	Dodecahedron	CuO, 197(4), 279 pm(4); $\theta_a = 28.5°$; $\theta_b = 89°$

coordination (Table 11.39). The factors influencing the formation of eight coordination and the stereochemical arrangement will be discussed.

Metal Influence

Size. Clearly, the metal ion radius needs to be large if one is to place eight donor atoms around it so that ligand-ligand repulsions are compatible with the metal-ligand bond strength. The largest metal ions are those on the left of the transition metal series (cf. Table 11.40). The line through the table divides the elements into those that form eight-coordinate compounds and those that do not. The demarcation is not sharp as regards radius size, indicating that although important, size is not the only factor in determining the coordination number.

Table 11.39 Some Eight-Coordinate Compounds

Ti, Zr, Hf

$Hf(NO_3)_4 4H_2O$	$Ti(bipy)_2Cl_4$
$Ti(S_2CNEt_2)_4$ (Zr and Hf)	$ZrBr_4(HNRR^1)_4$
$[Ti(CN)_8]^{5-}$	$Zr(tropolinate)_4$ (and Hf)
$Zr(BH_4)_4$ (and Hf)	
$Zr(CF_3CO_2)_4$	

V, Nb, Ta

$[Nb(tropolinate)_4]^+$ (and Ta),	$Nb(S_2CNMe_2)_4$

Cr, Mo, W

$[Mo(O_2)_4]^{2-}$ (and W)	$[Mo(OH)_2(CN)_5NO]^{4-}$
$[Mo(CN)_4(OH)_4]^{4-}$ (and W)	$[WBr_3(R_3N)_5]^{2+}$
$[Mo(CO_3)_4]^{4-}$	(R aromatic)
$[Mo(C_2O_4)_4]^{4-}$	

Mn, Tc, Re

$Mn(LL)_4$, $LL =$ ... or ...

$[TcCl_4(diars)_2]^+$	$[Re(CN)_8]^{2-,3-}$
$[ReX_4(diars)_2]^+$ ($X = Cl, Br$)	$[Re(CN)_4(OH)_4]^{4-}$
$[Re(CN)_7NO]^{3-}$	

Others

$[Fe(NO_3)_4]^-$, $Pd(NO_3)_4$, $[M(NO_3)_4]^{2-}$, cation Ph_3MeAs^+ ($M = Co, Ni, Cu$)

Table 11.40 Metallic and Covalent Radius (pm)*

Ti	V	Cr	Mn	Fe
147	135	129	137	126
132	*122*	*117*	*117*	*117*
Zr	Nb	Mo	Tc	Ru
160	147	140	135	134
145	*134*	*129*	*(127)*	*124*
Hf	Ta	W	Re	Os
159	147	141	137	135
144	*134*	*130*	*128*	*126*
	8 coordination		< 8 coordination	

* Metallic radius (roman) and *Covalent radius* (italic).

Oxidation State and Charge. Both a high oxidation state and the corresponding high positive charge on the metal ion favor eight coordination, since to maintain electroneutrality in a complex containing eight donor atoms the positive charge on the metal has to be reasonably high. However, it has been suggested that if the charge is too high, excessive polarization of the ligand may occur, which will lead to increased ligand-ligand repulsions.

Available Orbitals. As eight-bonding orbitals are necessary, metal ions with low d-electron configurations, d^0, d^1, and d^2 spin paired, are the most favored.

Energy of Bonding Orbitals. The energy separation between the bonding $(n - 1)d$, ns, and np orbitals increases as the atomic number increases, making it more difficult to achieve high coordination numbers as one moves to the right of the transition series. Therefore, orbital energy differences favor eight coordination on the left side.

Table 11.41   Eight Coordination

Group	d^9	d^7	d^4	d^3	d^2	d^1	d^0
IV							Ti(IV) Zr(IV) Hf(IV)
V				Nb clusters Ta clusters		V(IV) Nb(IV) Ta(IV)	V(V) Nb(V) Ta(V)
VI				Mo(III)	Mo(IV) W(IV)	Cr(V) Mo(V) W(V)	Mo(VI) W(VI)
VII					Tc(V) Re(V)	Re(VI)	Re(VII)
VIII			Ru(IV)				
		Co(II)					
	Cu(II)						

The transition metals that form eight-coordination compounds are given in Table 11.41, and are the metal ions predicted from the preceding considerations.

Ligand Influence

Electronegative and Nonpolarizable. The main type of donor atom associated with eight coordination is the highly electronegative and nonpolarizable ligand F^-. Both these properties enable high coordination numbers to be obtained while electroneutrality is maintained.

Size. Whereas the metal needs to be large, the ligand donor atoms should be small. This is the case for F^- and oxygen donor ligands, as well as bidentates with a small chelate span, such as in O_2^{2-}, 150 pm; NO_3^-, 210 pm; and RCO_2^-, 220 pm.

Counter-Ion Effect. The size of the counter ion can also affect the coordination number. For example, the complex salts Na_3TaF_8, K_2TaF_7, and $CsTaF_6$ can be obtained by adding the appropriate cation to the same solution. Presumably, factors such as crystal packing and relative sizes of the counter ions are important in this regard.

Crystal Field Stabilization Energy. Estimates of the CFSE for the two eight-coordinate shapes—square antiprism and dodecahedron—indicate that for low d-electron configurations the stabilization is greater than for six-coordinate octahedral complexes, by approximately $3Dq$.

Polarizable and π-Bonding Ligands. Ligands such as Cl^-, CN^-, and P and As donors form eight-coordinate compounds. This is rather unexpected in terms of what has been said heretofore. Owing to steric factors, the $M-L$ bond may be longer than normal, and so the ligand is less polarized. In addition, π bonding $M \rightarrow L$ or $L \rightarrow M$ may account for the complex stability with these ligands.

Stereochemistry

Of the various shapes possible for eight coordination, only two occur for transition metals, viz., the dodecahedron and square antiprism, both of which produce the least ligand-ligand repulsions. Both shapes can be considered to be derived from a cube by different distortions (Fig. 11.29).

Square Antiprism D_{4d}. The eight points of a square antiprism are equivalent, and the LML bond angle is 57–60°. The metal bonding orbitals used in an antiprismatic shape are d^4sp^3, the unused orbital being the d_{z^2}. All the bonding theories (VB, MO, and LF) agree that the d_{z^2} orbital is the most stable

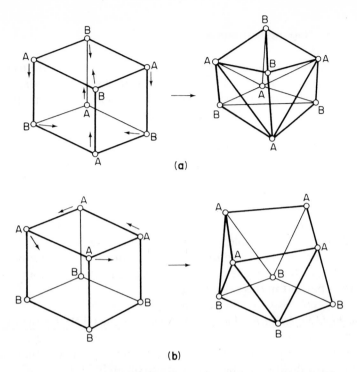

Fig. 11.29   The relationship of a cube to (a) dodecahedron, (b) square antiprism.

[Fig. 11.30(a)]. This orbital will be vacant for the d^0 configuration or will have one or two electrons for the d^1 and d^2 configurations, respectively. The symmetry of the orbital (A_1) is such that it may participate in π bonding to some or all of the eight ligands. The bonding will be $M \rightarrow L$ or $L \rightarrow M$, depending on the d-electron configuration and ligand type.

Dodecahedron D_{2d}. The coordination points around a dodecahedron fall into two groups—A and B (Fig. 11.29). Positions A are the points of an elongated tetrahedron, whereas the positions B correspond to the points of an interpenetrating flattened tetrahedron. The special features of a dodecahedron are the bond angles AMA (θ_a) = 35–37° and BMB (θ_b) = 69–73°, and the bond lengths r_A and r_B. The experimental bond angles agree with these figures, although the range is greater, i.e., θ_a = 35–38° and θ_b = 65–81, 86°. The orbitals best suited to bond in a dodecahedron are d^4sp^3, the unused and most stable orbital being the $d_{x^2-y^2}$ [Fig. 11.30(b)]. The $d_{x^2-y^2}$ orbital may participate in π bonding with the ligands near the xy plane, i.e., type B (Fig. 11.29). This may account for the π-donor ligands being in the positions B in the Ti(IV) d^0 complex [TiCl$_4$(diarsine)$_2$].

Fig. 11.30 Orbital splitting patterns for (a) square antiprism, (b) dodecahedron.

Square Antiprism vs. Dodecahedron. The majority of complexes studied so far are dodecahedral (Table 11.38). There are no obvious features indicating which ligands favor one or another of the stereochemistries. The following tentative suggestions may be put forward.

(a) Complexes containing mixed ligands tend to favor the dodecahedron, presumably because there are two types of ligand sites.

(b) Because of the less rigid requirements of the dodecahedron, this stereochemistry may be favored by bidentate ligands. This is generally true but there are exceptions.

(c) From overlap considerations, the π-bonding ability of the d_{z^2} and $d_{x^2-y^2}$ orbitals in the antiprism and dodecahedron, respectively, is calculated to be in the rate $1:1.40$. If π bonding is important, then ligands forming π bonds (e.g., CN^-, Cl^-, As, and P donors) would favor the dodecahedron. The antiprismatic structures studied so far do not involve π-bonding ligands.

The octacyanides of molybdenum and tungsten are interesting. Table 11.42 gives the results of structural studies. Considerable confusion

Table 11.42 Octacyanides of Molybdenum and Tungsten

	Cyanide species	Solid state structure	Solution structure
d^2	$[Mo(CN)_8]^{4-}$ $[W(CN)_8]^{4-}$	Dodecahedral (X ray)	Dodecahedral Structure retained (from vibrational spectroscopy)
	$[W(CN)_8]^{4-}$ (crystalline acid)	Antiprism (X ray)	
d^1	$[Mo(CN)_8]^{3-}$ $[W(CN)_8]^{3-}$	Dodecahedral In $[Mo(CN)_8]^{4-}$ host lattice (ESR)	Antiprismatic (ESR)

exists over the structures in solution, and it appears that the experimental results do not allow an unambiguous assignment of structure to be made. Doubt still exists over the solid state structure of the Mo(V) and W(V) complexes, and the dodecahedral shape may be the result of the constraint of the host lattice.

Cobalt and Copper Compounds

The eight-coordinate compounds $[Co(NO_3)_4]^{2-}$, $[Co(CF_3CO_2)_4]^{2-}$, and $[Cu(CH_3CO_2)_4]^{2-}$ raise special problems. The data given in Table 11.38 indicate that the complexes have four normal $M-O$ bonds and four long $M-O$ bonds, suggesting that the compounds may be equally well described as having a tetrahedral stereochemistry. However, the cobalt tetranitrate complex ion does not have an electronic spectrum typical of a tetrahedral species. This stereochemical problem does not arise for the analogous titanium compound $Ti(NO_3)_4$, where all eight $Ti-O$ bonds are the same length.

11.8 NINE COORDINATION

Nine coordination is rare for transition metals, and is so far only known for a few species, viz., the hydrides of rhenium(VII) and technetium(VII) and metal halide cluster compounds of niobium, tantalum, molybdenum, and tungsten.

The hydrides have the expected stereochemistry of a tripyramid (Fig. 11.31). The $Re-H_{av}$ bond is 168 pm and nmr studies show that the protons are all equivalent, indicating rapid exchange by a $M-H$ deformation

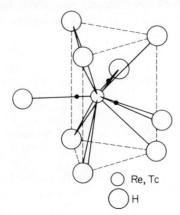

○ Re, Tc

○ H

Fig. 11.31 Tripyramidal structure of $[ReH_9]^{2-}$ and $[TcH_9]^{2-}$.

vibration. All nine of the metal valence orbitals are involved in the bonding in the complex ion.

The binary halides of molybdenum(II), tungsten(II), and the hexanuclear cluster halides of niobium and tantalum, as well as their complexes, normally contain the metal ions in ninefold coordination (Fig. 11.32). Eight of the

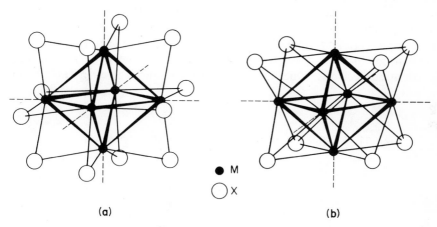

Fig. 11.32 Metal halide cluster species (a) M_6X_{12} and (b) M_6X_8.

bonds are involved in the cluster entity M_6X_8 or M_6X_{12}: four to neighboring metal ions and four to cluster halogens. The bond arrangement around each metal in M_6X_8 is that of a very distorted square antiprism, whereas in M_6X_{12} it is that of a distorted cube. The ninth bond of each metal is directed centrifugally from the faces of the clusters (dotted lines in Fig. 11.32).

Ninefold coordination is therefore achieved in two ways: (1) by using all nine orbitals of a metal ion with the d^0 configuration; clearly ligand size and polarization are important here; (2) by using all nine orbitals for a metal ion with the d^n configuration ($n \neq 0$), where the metal d electrons are involved in extensive metal-metal bonding. The limitation to an extension of the latter process is the total charge that becomes associated with the basic cluster unit. The requirements are for n to be small ($n \leqslant 4$) and the metal oxidation state to be low.

11.9 TRANSITION METAL π COMPLEXES AND CARBONYLS

A major field of transition metal chemistry not mentioned so far is that of organometallic π complexes and transition metal carbonyls. Except for the simplest of compounds, π complexes are not readily described in terms of traditional stereochemical approaches. For example, ferrocene (Fig. 11.33) may be considered to contain two-coordinate iron or six-coordinate iron, if it is assumed that in the latter case each cyclopentadienyl ring fills three coordination sites. Both descriptions are rather arbitrary.

The dominant characteristic of these compounds (and carbonyls) is the marked tendency to achieve the inert gas configuration. This involves filling up the valence $(n - 1)d$, ns, and np orbitals with 18 electrons. Hence, the coordination number of the metal, the stoichiometry, and the shape of the complex are largely determined by the 18-electron requirement of the metal atom. In using the 18-electron rule, it is necessary to formulate rules for counting electrons. One self-consistent set of rules is given in Table 11.43.

Table 11.43   Rules for Formulating Organometallic Compounds

1. Consider metal atom in zero-valent state.
2. Consider all ligands as neutral, donating, electrons, as follows:

Electron Donors	*Example*
1	CH_3, Cl, H, etc.
2	CO, P, As donors, bridging halogen $M-X \rightarrow$, C_2H_4 (i.e., double bond), etc.
3	allyl, NO
4	conjugated diene, e.g., butadiene, cyclobutadiene
5	dienyl ligands, e.g., cyclopentadienyl
6	triene, arene ligands, e.g., benzene
7	trienyl ligands, e.g., cycloheptatrienyl
8	cyclooctatetraene

3. Metal-metal bonds involve one electron from each metal.

A simple demonstration of the role of the inert gas configuration in determining the coordination number and shape of a complex may be seen from the metal carbonyls examples given in Table 8.3.

Table 11.44 lists a number of carbonyl and transition metal π complexes in which it is clear that the inert gas configuration is being achieved. The coordination numbers of the metal atoms in the complexes are also listed. One formal way of obtaining the coordination number is to equate the number of coordination sites occupied by the π ligand with $n/2$ (n even) or $n/2 + 1/2$ (n odd) for $n = 1$ to 6, where n is the formal number of electrons donated to the metal by the ligand (cf. Table 11.43). When $n = 7$ or 8, it is still best to consider the coordination number as 3 rather than 4. Using this method, we see that high coordination numbers are relatively common. However, we still cannot readily say what the coordination number is in certain cases. For example, butadiene can be considered to bond in one of two ways:

filling two or three coordination sites, respectively.

Also, little or no prediction can be made on the structure of the π complexes and carbonyl complexes on the basis of stoichiometry. For example, the reaction of $Fe_3(CO)_{12}$ with acetylene gives two isomeric forms of $[Fe_2(CO)_6(C_2H_2)_3]$. The structures of both isomers have been determined (Table 11.44), and it is apparent that both would have been difficult to predict. Mass spectral data can give useful structural information, but the definitive evidence comes from X-ray structural work, and as more structures are determined a new stereochemical pattern may emerge. Apparently, we shall need to put aside preconceived ideas on the directional properties of atomic orbitals in describing such complexes. Although such ideas are very useful in predicting, for example, the stereochemistry of the Co atom in $[Co(NH_3)_6]^{3+}$, they are of little help in predicting the stereochemistry of Co in a complex such as $Co_4(CO)_{10}(C_2H_5C{\equiv}CC_2H_5)$.

Square Planar Complexes

A number of π complexes of Pd(II), Pt(II), and, in some cases, Ni(II) have 16 valence electrons rather than 18 (Table 11.45). The compounds have another distinguishing feature—i.e., that the stereochemistry around the metal is based on the square plane. The reason for 16 valence electrons is probably that, as in other square planar compounds of metal atoms with the d^8 configuration, the p_z orbital perpendicular to the square plane is still vacant.

The metal halide cluster complexes discussed earlier (p. 259) are in many ways similar to the metal carbonyls, in that the 18-electron rule is achieved (or nearly so for Nb and Ta) by extensive metal-metal bonding. In fact, there is a marked division between the metal atoms that form halide and carbonyl clusters (Table 11.46). However, such a division may become more diffuse as further investigations are carried out on metal-metal bonded clusters. The division is partly due to the ligand types. The π-acceptor carbonyl ligands favor metal atoms with a lot of d electrons, whereas the opposite holds for halide ligands.

Table 11.44 Transition Metal Organometallic Compounds

Compound	Valence electrons		Coordination Number	Structure
$Mo(\pi\text{-}Cp)_2H_2$	Mo	6	8	
	2Cp	10		
	2H	2		
$Rh(\pi\text{-}Cp)[(CF_3)_6C_6]$	Rh	9	6	
	Cp	5		
	2C	2		
	‖ →	2		
$Co_2(CO)_6(PhC\equiv CPh)$	Co' and Co"		6	Octahedral?
	Co	9		
	Co—Co	1		
	3CO	6		
	2(Co—C)	2		

262

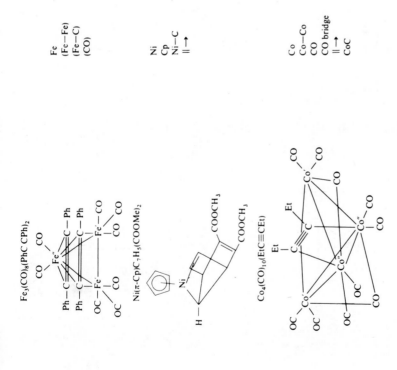

$Fe_3(CO)_8(PhC\ CPh)_2$

$Ni(\pi\text{-}Cp)C_7H_5(COOMe)_2$

$Co_4(CO)_{10}(EtC{\equiv}CEt)$

	Fe'	Fe"		
Fe	8	8	Fe'	7
(Fe—Fe)	2	2	Fe"	8
(Fe—C)	2	4		
(CO)	6	4		

Ni		
Ni	10	
Cp	5	5
Ni—C	1	
Ni=↑	2	

	Co'	Co"		
Co	9	9	Co'	6
Co—Co	2	3	Co"	7
CO	4	4		
CO bridge	1	1		
=↑ CoC	2	1		

Table 11.44 (cont.)

Compound	Valence electrons			Coordination number		Structure
Fe$_2$(CO)$_6$(C$_2$H$_2$)$_3$	Fe 8					
	CO 6					
	FeC 1			6		Octahedral?
	=↑ 2					
	Fe—Fe 1					
		Fe'	Fe"			
	Fe	8	8	Fe'	6	
	CO	4	8	Fe"	6	Fe" Octahedral
	Cp	5				
	Fe—C	1	1			
	FeFe	1	1			
Ni(C$_6$Me$_4$O$_2$)C$_8$H$_{12}$	Ni 10			4		Tetrahedral
	=↑ 8					
Fe(Ph$_4$C$_4$)(CO)$_3$	Fe 8					
	CO 6			5 or 6?		
	▭ 4					

Re(π-Cp)(CH₃)₂(C₅H₅Me)

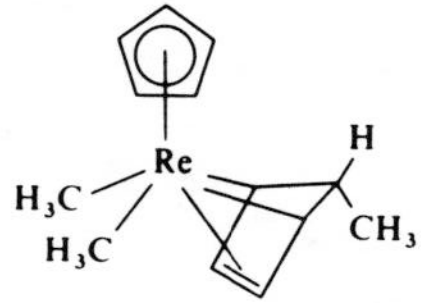

Re	7
Cp	5
Re—C $\rightleftharpoons$	4
$\|$	2

8

Mn₂(CO)₄(diars)(π-MeC₅H₄)₂

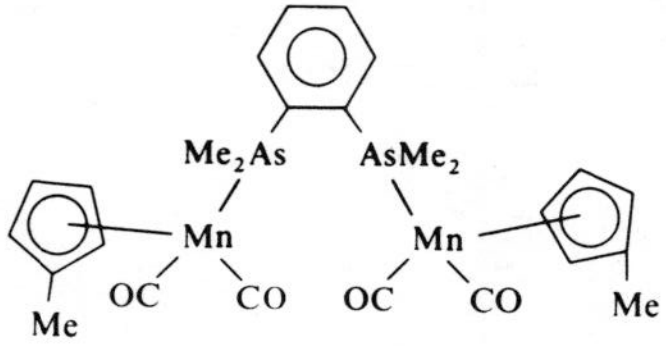

Mn	7
Cp	5
CO	4
As	2

6

Ir(C₇H₈)₂SnCl₃

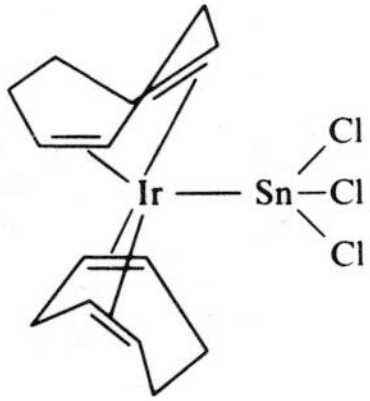

Ir	9
IrSn	1
$\|$ $\rightleftharpoons$	8

5   Square pyramid

V(π-C₇H₇)(CO)₃

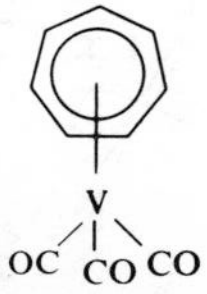

V	5
CO	6
C₇H₇	7

6?

Table 11.45 Square Planar Organometallic Compounds

Compound	Valence electrons		Coordination number
$Pd_2Cl_2(C_3H_5)_2$			
	Pd	10	
	Cl	1	4 or 3
	Cl_{br}	2	
	⟨ →	3	
$[Pd(C_3H_5)(acetate)]_2$			
	Pd	10	
	O^-	1	4 or 3
	O→	2	
	⟨ →	3	
$PdClPPh_3CH_3C(CH_2)_2$			
	Pd	10	
	Ph_3P	2	
	Cl	1	4
	PdC	1	
	‖ →	2	
$Ni[CH_3C(CH_2)_2]_2$			
	Ni	10	
	⟨ →	6	4 or 2
$Pd(acac)(C_8H_{11})$			
	Pd	10	
	O^-	1	4 or 3
	O→	2	
	⟨ →	3	
$[PtCl_2C_2H_4]_2$			
	Pt	10	
	Cl	2	4
	Cl_b	2	
	‖ →	2	

Table 11.46 Transition Metals Forming Metal Clusters

		Mn	Fe	Co
Nb	Mo	Tc	Ru	Rh
Ta	W	Re	Os	Ir
Metal halide clusters			Metal carbonyl clusters	
Oxidation state 2, 3, (4)			Oxidation state normally zero	

11.10 TEMPLATE REACTIONS

Since ligands are fixed in certain positions in space around a metal ion, it is possible to make use of this fact to bring about a chemical reaction dependent on the positioning of certain reactive sites. Such reactions are termed "template." In such systems, two effects may be operating in the reaction of coordinated ligands: (1) the template effect; and (2) the electronic effect of the metal ion. We are only concerned with the first effect at present. The reactions discussed previously (p. 239) on homogeneous catalysis are examples of template reactions where the organic ligand is positioned to allow a reaction such as hydrogenation or polymerization to occur. The synthesis of phthalocyanines is assisted by metal ions, and it is likely that the metal ions provide a reactive template site. A few examples of template reactions are now given. Acetone reacts with $[Ni(en)_3]^{2+}$ to give a quadridentate ligand:

$$[Ni(en)_3]^{2+} + CH_3COCH_3 \rightarrow$$

The reaction of α diketones and β mercapto-amines to give a Schiff base is not possible, because the amino and mercapto groups are competitive nucleophiles. However, if a metal ion is introduced, the required reaction occurs because the sulfurs are now tied to the nickel atom.

When $R = CH_3$, this complex can be reacted further by taking advantage of the two sulfur atoms in the cis position

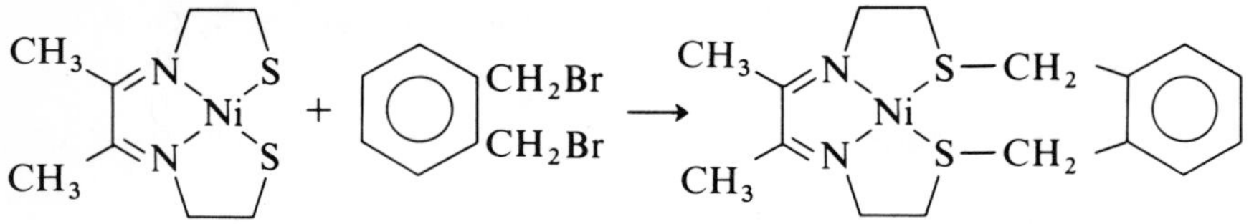

Sexadentate ligands can also be produced by using the template effect:

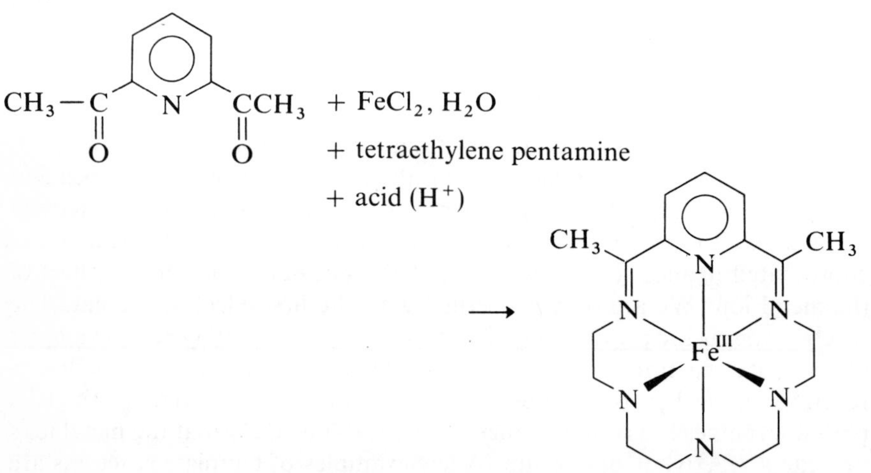

$CH_3-C \quad N \quad CCH_3$ + $FeCl_2$, H_2O

+ tetraethylene pentamine

+ acid (H^+)

$\longrightarrow$

If triethylene tetramine is used, a quinquidentate ligand is produced.
A final example is given from the chemistry of palladium:

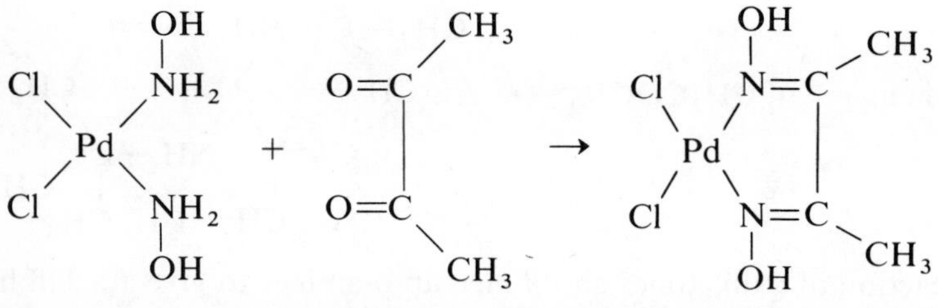

One feature common to most of these reactions is the placing of certain reactive sites cis to each other in the complex, allowing the gap to be closed:

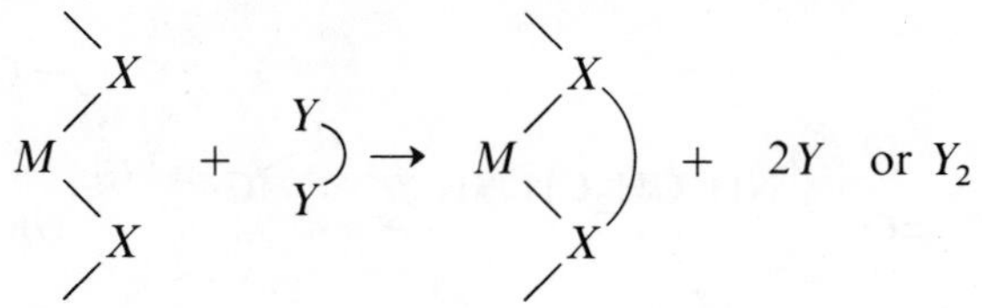

It is very likely that many template reactions have been used in chemical reactions without being recognized as such. Finally, deliberate use of the template effect can lead to new ligands not necessarily available by any other reaction.

BIBLIOGRAPHY

1. H. B. GRAY and C. J. BALLHAUSEN, *J. Am. Chem. Soc.*, **85**, 260 (1963).

2. H. B. GRAY, *Transition Metal Chemistry*, Vol. 1, ed. by R. L. Carlin, Marcel Dekker, Inc., New York, 1965, 240.

3. F. BASOLO and R. C. PEARSON, *Mechanism of Inorganic Reactions*, 3rd ed., John Wiley & Sons, Inc., New York, 1967.

4. R. S. NYHOLM, *Proc. Chem. Soc.*, 273 (1961).

5. F. A. COTTON and G. WILKINSON, *Advanced Inorganic Chemistry*, 3rd ed., Interscience Publishers, New York, 1972.

6. R. J. GILLESPIE, *J. Chem. Soc.*, 4679 (1963).

7. E. L. MUETTERTIES and R. A. SCHUNN, *Quart. Rev. London*, **20**, 245 (1966).

8. C. FURLANI, *Coord. Chem. Rev.*, **3**, 141 (1968).

9. L. E. SUTTON, ed., *Interatomic Distances, Chem. Soc. London Spec. Publ.* 11 (1958).

10. L. E. SUTTON, ed., *Interatomic Distances, Chem. Soc. London Suppl. Spec. Publ.* 18 (1965).

11. J. A. McCLEVERTY, *Prog. Inorg. Chem.*, **10**, 49 (1968).

12. S. J. LIPPARD, *Progr. Inorg. Chem.*, **8**, 109 (1967).

13. R. V. PARISH, *Coord. Chem. Rev.*, **1**, 439 (1966).

FURTHER READING

J. D. DUNITZ and L. E. ORGEL, *Advan. Inorg. Chem. Radiochem.*, **2**, 1 (1960).

R. EISENBERG, *Progr. Inorg. Chem.*, **12**, 295 (1970).

P. J. WHEATLEY, *Perspectives Struct. Chem.*, **1**, 1 (1967).

M. C. BAIRD, *Progr. Inorg. Chem.*, **9**, 1 (1968).

B. R. PENFOLD, *Perspectives Struct. Chem.*, **2**, 71 (1968).

12

The Lanthanides and Actinides

The two series of elements, the lanthanides and actinides, stand apart from the other elements of the periodic table in a number of ways. First, the outer electron structure is

$$nd^{10}nf^{x}(n + 1)d^{0 \text{ or } 1}(n + 1)s^2$$

where $n = 4$ for the lanthanides and 5 for the actinides, and the value of x ranges from 0 to 14. Second, the coordination numbers of the metal ions in compounds are high, reaching 12 for the actinides and hardly, if ever, dropping below 6 for either series (at least in the solid state). Third, the metallic and ionic radii of the elements are large compared with the size of the transition metals (Table 12.1).

12.1 STEREOCHEMISTRY OF THE LANTHANIDES AND ACTINIDES

f Orbitals

The 4f and 5f orbitals are progressively being filled for the lanthanides and actinides, respectively. For the lanthanides, the 4f orbitals are deep-seated and unlikely to be involved in bonding. But in the case of

Table 12.1   Atomic and Ionic Radii of the Transition Metals, Lanthanides, and Actinides

Atomic radius (pm)

Third transition metal series

Hf	Ta	W	Re	Os	Ir	Pt
159	147	141	137	135	136	139

Lanthanides

La	Ce	Pr	Nd	Pm	Sm	Eu	Gd	Tb	Dy	Ho	Er	Tm	Yb	Lu
188	181	182	182			206	179	177	177	176	175	174	194	172

Actinides

Th	Pa	U	Np	Pu	Am
180	163	156	156	163	182

Ionic radius (pm)

Third transition metal series

Hf^{4+}	Ta^{5+}	W^{4+}	Re^{7+}	Os^{4+}	Ir^{4+}	Pt^{2+}
75	73	68	68	67	66	52

Lanthanides M^{3+}

La	Ce	Pr	Nd	Pm	Sm	Eu	Gd	Tb	Dy	Ho	Er	Tm	Yb	Lu
106	103	101	101	98	96	95	94	92	91	89	88	87	86	85

Actinides

M^{3+}

Ac	Th	Pa	U	Np	Pu	Am	Cm
111	(108)	(105)	103	101	100	99	98

M^{4+}

Th	Pa	U	Np
83	82	81	80

the actinides, the $5f$ orbitals are nearer the surface of the atom and may be involved in bonding. However, it is probably not meaningful to consider stereochemical configurations in terms of the involvement of f orbitals in bonding.

Ionic Compounds

Since the f orbitals are best considered as nonbonding, it is probably more correct to describe lanthanide and actinide compounds as ionic assemblages. The high coordination numbers can therefore be related to the large ionic radii of the elements (Table 12.1). However, attempts to arrive at coordination numbers by use of the radius ratio rule have not been very successful. For simple binary lanthanide complexes, there is a trend of decrease in coordination number with fall in ionic radius (Table 12.2). The

Table 12.2   Stereochemistry of Some Lanthanide Compounds

	MF_3	MCl_3	MBr_3	MI_3	$M(OH)_3$	M_2O_3	MOF	MOCl
La	$5 + 6^a$	9^c	9^c	8^f	9^c	7^f	8^i	8^j
Ce	$5 + 6^a$	9^c	9^c			$7^f, 6^h$	8^i	8^j
Pr	$5 + 6^a$	9^c	9^c		9^c	$7^{f,g}, 6^h$	8^i	8^j
Nd	$5 + 6^a$	9^c	8^e		9^c	$7^{f,g}, 6^h$	8^i	8^j
Pm		9^c						
Sm	$\begin{cases} 5 + 6^a \\ 8 + 1^b \end{cases}$	9^c	8^e		9^c	$7^{f,g}, 6^h$	8^i	8^j
Eu	$\begin{cases} 5 + 6^a \\ 8 + 1^b \end{cases}$					$7^g, 6^h$	8^i	8^j
Gd	$8 + 1^b$				9^c	$7^g, 6^h$	8^i	8^j
Tb	$8 + 1^b$					6^h	8^i	8^j
Dy	$8 + 1^b$	6^d			9^c	$7^g, 6^h$		8^j
Ho	$\begin{cases} 5 + 6^a \\ 8 + 1^b \end{cases}$	6^d				6^h	8^i	8^j
Er	$8 + 1^b$	6^d			9^c	6^h		8^j
Tm	$\begin{cases} 5 + 6^a \\ 8 + 1^b \end{cases}$	6^d				6^h		
Yb	$8 + 1^b$	6^d				6^h		
Lu	$8 + 1^b$	6^d				6^h		

[a] Trigonal bipyramid + trigonal prism.
[b] Tripyramid (one bond longer than rest).
[c] UCl_3 structure tripyramid (Fig. 12.1).
[d] $AlCl_3$ structure (Fig. 11.25c).
[e] $PuBr_3$ structure (Fig. 12.2).
[f] Pentagonal bipyramid.
[g] Trigonal prism + one.
[h] Cube—two corners missing.
[i] CaF_2 structure (Fig. 12.3).
[j] BiOCl structure (Fig. 12.4).

ionic character of the compounds is also suggested by the type of species formed. Ligands that bond to the metals are the anionic type (halogen, oxygen, acetylacetone) or neutral ligands (such as water) that readily form ion-dipole interactions. But neutral ligands requiring a high proportion of covalent bonding (such as Ph_3P, CO) do not coordinate.

Stereochemistry

Tables 12.2 through 12.5 list some compounds of the lanthanide and actinides having known structures. It is evident that in addition to the high coordination numbers there is a wide range of shapes for each coordination number. For example, the usual square antiprismatic and dodecahedral

Table 12.3 High Coordination Numbers for the Lanthanide Elements

Coordination number	Compound	Structure
10	$[La(H_2O)_4(EDTA)]3H_2O$	
9	$K[La(H_2O)_3(EDTA)]5H_2O$	
	$NaNdF_4$	
	$NaLaF_4$ (and K^+)	
	$NaCeF_4$ (and K^+)	
8	$[GdCl_2 6H_2O]Cl$	Square antiprism
	also Nd, Sm, Eu	
	$Ce(acac)_4$	Square antiprism
	$[Eu(thenoyltrifluoroacetylacetone)_3(H_2O)_2]$	Square antiprism
	$[Y(acac)_3(H_2O)_2]H_2O$	Square antiprism
	$Ce(IO_3)_4$	Square antiprism
	$Ce(IO_3)_4 H_2O$	Square antiprism
	$CeO(SO_4)H_2O$	Square antiprism
	$Ce(dibenzoylmethane)_4$	Dodecahedron
6	$(pyH)_2CeCl_6$	Octahedron

shapes occur for eight-coordinate systems but so does the hexagonal bipyramid. The predominance of this latter shape for uranium is probably a result of the stable linear entity UO_2^{2+} in a number of compounds.

The change in structure for the lanthanide trifluorides and sesquioxides, i.e., trigonal prism + trigonal bipyramid to tripyramid and pentagonal bipyramid to trigonal prism + one additional bond, is probably related to the change in ionic radius of the metal ion. As the radius gets smaller, the structure changes to a less crowded one.

Table 12.4 Stereochemistry of Some Actinide Compounds

	MF₃	MCl₃	MBr₃	MI₃	MOX	MF₄	MCl₄	MBr₄	M₂O₃	MO₂	MF₅	MCl₅	MF₆	MCl₆
Ac	5 + 6[a]	9[b]	9[b]		8[d]				7[f]					
Th						4 + 4[e]	4 + 4[e]	4 + 4[e]		8[h]				
Pa						4 + 4[e]	4 + 4[e]			8[h]	7[j]	7[j]		
U	5 + 6[a]	9[b]	9[b]	8[c]		4 + 4[e]	4 + 4[e]			8[h]	6[i],7[j]	6[i]	6[i]	6[i]
Np	5 + 6[a]	9[b]	9[b], 8[c]	8[c]	8[d]	4 + 4[e]	4 + 4[e]			8[h]			6[i]	
Pu	5 + 6[a]	9[b]	8[c]	8[c]	8[d]	4 + 4[e]			6[g]	8[h]				
Am	5 + 6[a]	9[b]	8[c]	8[c]	8[d]	4 + 4[e]			7[f], 6[g]	8[h]				
Cm									6[g]					
Bk					8[d]									
Cf		9[b]							7[k]					

[a] Trigonal prism + trigonal bipyramid LaF₃ type.
[b] UCl₃ structure (Fig. 12.1).
[c] PuBr₃ structure (Fig. 12.2).
[d] BiOCl structure (Fig. 12.4).
[e] Dodecahedral structure.
[f] Pentagonal bipyramid.
[g] Cube, two corners missing.
[h] CaF₂ structure (Fig. 12.3).
[i] Octahedral.
[j] βUF₅ structure (Fig. 12.5).
[k] Trigonal prism + one.

Table 12.5 High Coordination Numbers for the Actinide Elements

Coordination number	Compound	Structure
12	$[Th(NO_3)_6][Mg(H_2O)_6]$	Irregular icosahedron
11	$Th(NO_3)_4 5H_2O$	
10	$U(CH_3CO_2)_4$ polymer	
9	K_2PaF_7	
	$LiUF_5$	
	$\beta_2 Na_2UF_6$ ($\beta_1 K^+$)	Tripyramid
	$\beta_2 NaThF_6$	
	$\beta_2 NaPuF_4$	
	$\beta_2 K_2UF_6$	
	$\beta_1 K_2ThF_6$	
	$\beta_1 K_2UF_6$	
	U_2F_9	
	$NaTh_2F_9$	
	K_2AcF_7	
8	$Rb[UO_2(NO_3)_3]$	Hexagonal bipyramid
	$Na[MO_2(CH_3CO_2)_3]$ (M = U, Np, Pu, Am)	Hexagonal bipyramid
	$NH_4[UO_2(cupferron)_3]$	Hexagonal bipyramid
	$[UO_2(NO_3)_2(H_2O)_2]$ 4, 1, OH_2O	Hexagonal bipyramid
	$[UO_2(NO_3)_2(triethylphosphate)_2]$	Hexagonal bipyramid
	αUO_3	Hexagonal bipyramid
	UO_2F_2	Hexagonal bipyramid
	$Ca[UO_2(O_2)]$	Hexagonal bipyramid
	$\alpha UO_2(OH)_2$	Hexagonal bipyramid
	$UO_2(CO_3)$	Hexagonal bipyramid
	$K[MO_2CO_3]$ (M = U, Pu, Am)	Hexagonal bipyramid
	U_3O_8	Hexagonal bipyramid
	Na_2UF_6	
	$\alpha M(acac)_4$ (M = U, Th)	Square antiprism
	$\beta M(acac)_4$ (M = Th, Pa)	Square antiprism
	$Pu(IO_3)_4$	Square antiprism
	$M(OH)_2SO_4$ (M = U, Th)	Square antiprism
	$Th(OH)_2CrO_4 H_2O$	Square antiprism
	$U_6O_4(OH)_4(SO_4)_6$	Square antiprism
	$Na_4[M(C_2O_4)_4]3H_2O$ (M = U, Th)	Dodecahedron
	$M(dibenzoylmethane)_4$ (M = U, Th)	Dodecahedron
7	$[(UO_2)_2(SO_4)_3]_n^{2n-}$	Pentagonal bipyramid
	K_3UF_7	Pentagonal bipyramid
	Na_3UF_7	Pentagonal bipyramid
	$K_3[UO_2F_5]$	Pentagonal bipyramid
6	$Cs_2[UO_2Cl_4]$	Octahedron
	$PuCl_6^{2-}$	Octahedron
	MF_6 (M = Pu, Np, U)	Octahedron

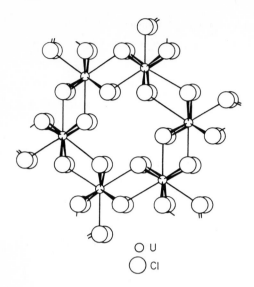

○ U

○ Cl

Fig. 12.1   Structure of UCl₃.

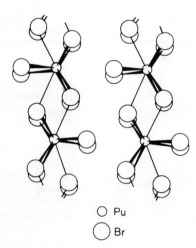

○ Pu

○ Br

Fig. 12.2   Structure of PuBr₃.

In a number of compounds (e.g., lanthanide and actinide trifluorides), there are distinct nearest and next-nearest neighbors, and the difference in distances is small enough to consider the coordination number to be equal to the sum of the nearest and next-nearest neighbors. For example, in LaF₃, the bond lengths are;

$$5 \text{ La—F} \quad 236 \text{ pm} \quad \text{(trigonal bipyramid)}$$

and

$$6 \text{ La—F} \quad 270 \text{ pm} \quad \text{(trigonal prism)}$$

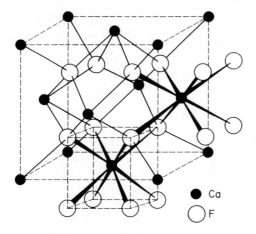

Fig. 12.3   CaF₂ structure.

giving an effective coordination number of 11. In YF_3, they are;

$$8 \ Y-F \qquad 230 \ pm$$

and

$$1 \ Y-F \qquad 260 \ pm$$

giving an effective coordination number of 9 (a distorted tripyramid). In the case of the MX_4 compounds of the actinides, the difference between the nearest and next-nearest neighbors is greater. For example, in UCl_4,

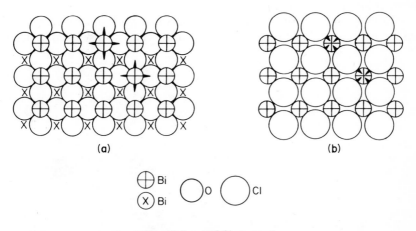

Fig. 12.4   BiOCl structure.

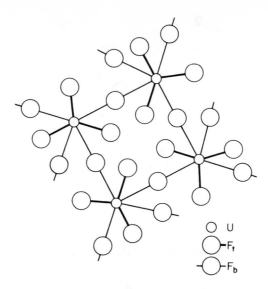

Fig. 12.5 Structure of βUF_5.

the bond lengths are:

$$4 \text{ U—Cl} \quad 241 \text{ pm}$$

and

$$4 \text{ U—Cl} \quad 309 \text{ pm}$$

This is thought to be the result of a compromise between some covalent and ionic bonding. The U—Cl ionic bond length is predicted to be $r_{U^{4+}} + r_{Cl^-} = 93 + 181 = 274$ pm, but this could be increased to 309 pm owing to chlorine (short)-chlorine (long) repulsions.

Considerably more chemical and structural work is necessary in this field before more general stereochemical features can be discussed in detail.

Appendix
Abbreviations Used for Ligands

The following is an alphabetical list of abbreviations used for ligands mentioned throughout the text and tables.

Abbreviation	*Name*	*Formula*
acac	acetylacetonate	
AHA	acetylacetone-mono-(o-hydroxyanilate)	

Abbreviation	*Name*	*Formula*
acac-pn	bisacetylacetonatopropane 1,3 diamine	
biby	αα′ bipyridyl	
π-Cp	π-bonded cyclopentadiene	
cpm	3,4′ dicarboxyethyl-5-chloro 3,4,5′ trimethyldipyrromethenate	
DAMPE	diacetylmonoxime-2-(2 pyridyl) ethylamine	
DEAS	4 chloro-N-2diethylaminoethyl salicyladiminate	
diars or diarsine	o-phenylenebisdimethylarsine	
diphos	1,2-bis(diphenylphosphino)-ethane	$Ph_2PCH_2CH_2PPh_2$
DMF	dimethylformamide	$(CH_3)_2NCOH$
DMG	dimethylglyoximate	

Abbreviation	*Name*	*Formula*
dpm	dipyrromethenate	
DTC	NN′dialkyldithiocarbamate	
EDTA	ethylenediaminetetraacetic acid	$(COOH)_2(CH_2)_2N(CH_2)_2N(CH_2)_2(COOH)_2$
Et_4dien	1,1,7,7 tetraethyldiethyl triamine	$HN[CH_2-CH_2-N(C_2H_5)_2]_2$
Me_6tren	tris(2 dimethylaminoethyl) amine	$N[CH_2CH_2-N(CH_3)_2]_3$
MNT	maleonitriledithiolate	
NTA	nitrilotriacetate	$N(CH_2COO)_3$
o-phen	o-phenanthroline	
PAPHY	pyridine-2-aldehyde-2′-pyridylhydrazone	
phth	phthalocyanine	
py	pyridine	

Abbreviation	*Name*	*Formula*
QAS	tris(*o*-diphenylarsinophenyl) arsine	$\left[\text{As}-\underset{}{\bigcirc}-\text{As}(C_6H_5)_2\right]_3$
QP	tris-(*o*-diphenylphosphino-phenyl) phosphine	$\left[\text{P}-\underset{}{\bigcirc}-\text{P}(C_6H_5)_2\right]_3$
Salicyl-en	NN′disalicylideneethylene diamine dianion	$\left[\begin{array}{c} \text{O} \\ \bigcirc\text{--}\underset{H}{\text{C}}\text{=N} \quad \text{N=CH--}\bigcirc\text{--O} \\ \text{CH}_2\text{--CH}_2 \end{array}\right]^{2-}$
Sal-Me or N-Me Sa	N methylsalicylaldiminato	$\left[\begin{array}{c} \text{O} \\ \bigcirc\text{--}\underset{H}{\text{C}}\text{=}\overset{N}{}\text{--CH}_3 \end{array}\right]^{-}$
Salicyl-pn	NN′ disalicylidene propane 1,2 diamine dianion	$\left[\begin{array}{c} \text{O} \\ \bigcirc\text{--CH=N} \quad \text{N=CH--}\bigcirc\text{--O} \\ \text{CH}_2\text{--CH} \\ \text{CH}_3 \end{array}\right]^{2-}$
TAP	tris(3-dimethylarsinopropyl) phosphine	$P[CH_2CH_2CH_2As(CH_3)_2]_3$
TAS	bis(*o*-diphenylarsinophenyl) phenyl arsine	$\left[C_6H_5\text{As}-\underset{}{\bigcirc}-\text{As}(C_6H_5)_2\right]_2$
TDT	toluene 3,4 dithiolate	$\left[CH_3\text{--}\bigcirc\underset{S}{\overset{S}{<}}\right]^{2-}$
terpy	αα′α″ terpyridyl	(three-ring pyridyl structure with N, N, N)
TMED	tetramethylethylene diamine	$(CH_3)_2NCH_2CH_2N(CH_3)_2$
tren	tris(2aminoethyl)amine	$N(CH_2CH_2NH_2)_3$

Abbreviation	Name	Formula
triars	methylbis(3dimethylarsino propyl)arsine	$CH_3As\begin{cases}(CH_2)_3As(CH_3)_2 \\ (CH_2)_3As(CH_3)_2\end{cases}$
triphos or TDPME	1,1,1 tris(diphenyl phosphinomethyl)ethane	$CH_3C(CH_2PPh_2)_3$
TSP	tris(*o*-methylthiophenyl) phosphine	$P(CH_3SC_6H_4)_3$

Index of Compounds

General Index